D1504654

Marine Geography

GIS for the Oceans and Seas

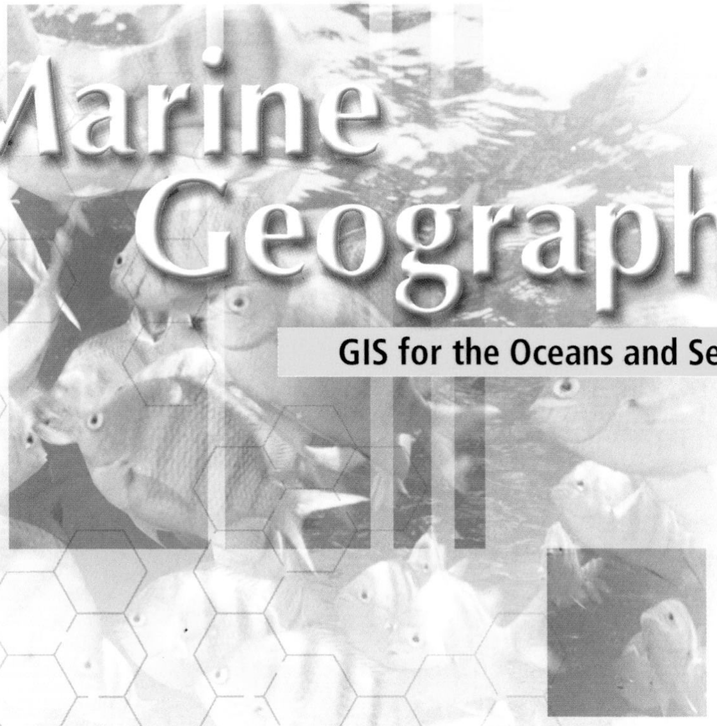

Edited by Joe Breman

ESRI Press
Redlands, California

Copyright © 2002 ESRI. All rights reserved.

The information contained in this document is the exclusive property of ESRI. This work is protected under United States copyright law and the copyright laws of the given countries of origin and applicable international laws, treaties, and/or conventions. No part of this work may be reproduced or transmitted in any form or by any means, electronic or mechanical, including photocopying or recording, or by any information storage or retrieval system, except as expressly permitted in writing by ESRI. All requests should be sent to Attention: Contracts Manager, ESRI, 380 New York Street, Redlands, California 92373-8100, USA.

The information contained in this document is subject to change without notice.

U.S. GOVERNMENT RESTRICTED/LIMITED RIGHTS: Any software, documentation, and/or data delivered hereunder is subject to the terms of the License Agreement. In no event shall the U.S. Government acquire greater than RESTRICTED/LIMITED RIGHTS. At a minimum, use, duplication, or disclosure by the U.S. Government is subject to restrictions as set forth in FAR §52.227-14 Alternates I, II, and III (JUN 1987); FAR §52.227-19 (JUN 1987) and/or FAR §12.211/12.212 (Commercial Technical Data/Computer Software); and DFARS §252.227-7015 (NOV 1995) (Technical Data) and/or DFARS §227.7202 (Computer Software), as applicable. Contractor/Manufacturer is ESRI, 380 New York Street, Redlands, California 92373-8100, USA.

ArcView, ESRI, ArcIMS, ARC/INFO, ArcGIS, ArcInfo, Avenue, AML, ArcSDE, Spatial Database Engine, ArcMap, ArcExplorer, 3D Analyst, ESRI, the ESRI Press logo, @esri.com, and www.esri.com are trademarks, registered trademarks, or service marks of ESRI in the United States, the European Community, or certain other jurisdictions. Microsoft and the Windows logo are registered trademarks and the Microsoft Internet Explorer logo is a trademark of Microsoft Corporation. Other companies and products mentioned herein are trademarks or registered trademarks of their respective trademark owners.

ESRI
Marine Geography: GIS for the Oceans and Seas
ISBN 1-58948-045-7

First printing October 2002.

Printed in the United States of America.

Library of Congress Cataloging-in-Publication Data
Marine geography : GIS for the oceans and seas / Joe Breman, editor.
p. cm.
Includes bibliographical references.
ISBN 1-58948-045-7 (pbk.)
1. Marine geographic information systems. 2. Oceanography—Remote sensing. I. Breman, Joe, 1970–
GC38.5.M37 2002
551.46—dc21 2002014582

Published by ESRI, 380 New York Street, Redlands, California 92373-8100.

Books from ESRI Press are available to resellers worldwide through Independent Publishers Group (IPG). For information on volume discounts, or to place an order, call IPG at 1-800-888-4741 in the United States, or at 312-337-0747 outside the United States.

Contents

Pacific

Northeast Pacific

Acknowledgments

I would like to recognize and thank all of the authors who contributed their time and energy toward the production of this book; it is their collaborative spirit that promotes growth in the field of marine GIS. I would also like to specifically recognize and thank Dawn Wright and Eric Treml for their peer review of the material and lending their scientific and academic expertise. My thanks also to David Boyles, for helping to edit the manuscript; Jennifer Johnston, for designing and producing the book; Edith M. Punt and Christian Harder, for editorial review and support; Tiffany Wilkerson and Michael Hyatt, for copyediting; Steve Pablo, for cover design; Barbara Shields, for review of initial stages of the work; Cliff Crabbe, for overseeing print production; and Jeanne Foust, for supervising marine initiatives at ESRI.

My thanks to Charles Convis, for his constant support, encouragement, and advice along the way, and Jack Dangermond, for the attention he has given to marine GIS and his philanthropic support of so many of the organizations and projects represented in this book.

Most of all I would like to give a heartfelt thanks to my wife, Galit, whose patience and understanding made this book possible.

Joe Breman, editor

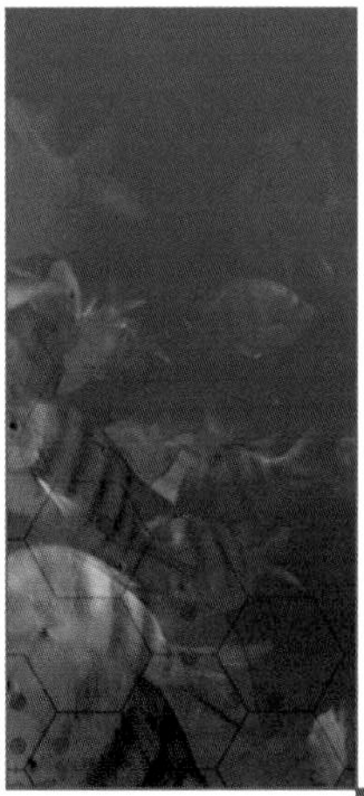

Foreword

Charles Convis
ESRI Conservation Program
www.conservationgis.org

As expanding human populations exert increasingly destructive pressures on the natural systems that sustain all life, the need for better conservation science tools becomes more and more acute. The biological and mathematical disciplines alone are inadequate to meet these challenges. Computer tools are also inadequate by themselves to solve these problems. What we need is a combined system of integrative theories, scientific procedures, and computer tools to bring the vast diversity of information into a manageable form. As conservationists, we think geographic information systems (GIS) technology provides a vital solution by integrating theoretical approaches from geography and ecology with powerful spatial database and statistical functions. In the same way that Newton's calculus allowed classical mechanics to become more predictive, we think GIS is the breakthrough tool that will allow the ecological sciences to become more predictive, rigorous, and directly integrated into all manner of social, political, and resource decisions.

To better support the use of these GIS tools among conservationists and biologists, ESRI has maintained a conservation support program since 1989. The ESRI Conservation Program has made in-kind grants to more than four thousand non-profit groups and has helped found two other independent support organizations, as well as specialized projects such as the new Marine Conservation Program. In 1998 we started "Conservation Geography," a series of magazines and books devoted to spatial tools in conservation science. In 2001, we conceived a special publication devoted to the uses of GIS in marine conservation.

This book presents a current snapshot of the progress of GIS in the marine sciences, showing how current theories in marine biology, geology, and conservation are being incorporated into GIS-based research systems, and how these systems are having important impacts on marine policy and coastal populations. It is our shared goal to support the conservation of the oceans and seas worldwide by facilitating communication and encouraging cooperation throughout the marine GIS user community.

Introduction

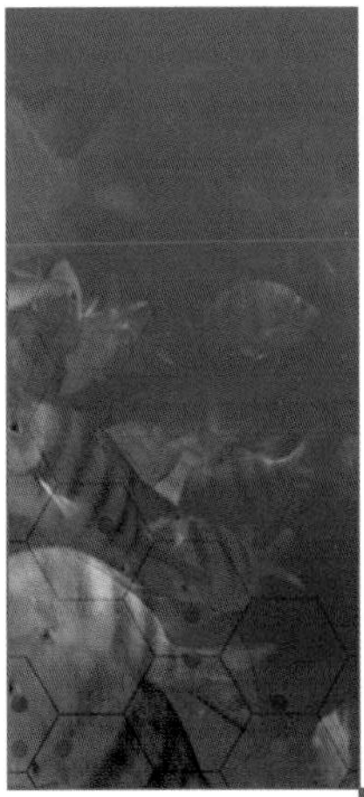

Joe Breman
Redlands, California

Recent publications and advancements in related research and development have helped to solidify the fast-growing discipline of marine GIS. In an effort to track the interest in this important niche, the first annual Marine Special Interest Group meeting was held at the ESRI User Conference in 2001. One of the most important goals expressed by this group was the need for the development of 3-D and 4-D tools that would make the modeling of oceans and associated marine layers more realistic, representative, and dynamic in nature. Also of primary importance to the group was a marine data model that would help steer the standardization and integration of marine GIS data collection and analysis. To answer this call, an initiative was launched in the subsequent months to develop a GIS marine data model. This work is ongoing. Once the preliminary feature classes and relationships of the data model are established, the real work begins: that is, loading data into the model, then testing and experimenting with the capabilities of the data structures and thinking "outside of the box" with regard to how we can work with dynamic marine data in a computer environment. The design and creation of the marine data model has been a collaborative process, and one that pointed to the many marine projects and studies in this discipline, emphasizing the growing need to document them.

In tandem with this important work, new connections and contacts were established among members of the marine community, and it became clear that there was a growing need to provide an outlet to share the collective progress of these studies. In his book, *Conservation Geography: Case Studies in GIS, Computer Mapping, and Activism* (published by ESRI Press in 2001), Charles Convis, director of the ESRI Conservation Program and a founding father of the Society for Conservation GIS (SCGIS), focused on a variety of projects that highlight conservation GIS. The number of marine-related works of this type warranted a division of its own, and a Marine Geography section evolved dedicated to conservation GIS work in the oceans and seas around the world.

Marine Geography: GIS for the Oceans and Seas is a collection of chapters submitted by researchers and experts who represent the cutting edge of the professional, academic, and nonprofit sectors of the global marine GIS community. At the present time, most GIS applications can create a graphic display of the seafloor or bathymetric layers; the sea surface is symbolized in most cases by shades of blue, and the coast by a simple line. Marine researchers and professionals are challenged to represent the many elements that are missing to present a true portrayal of the oceans and seas. Dawn Wright, oceanographic researcher and geography professor, has pointed out in her previous writings (as have other writers) that, with the technological advances of today, we may know more about Mars than we do about our own oceans and seas, and the biological, physical, and chemical complexities therein. The oceans and seas reflect many indicators of climate change, and rising sea levels are certain to become more and more important as they impinge on coastal population centers all over the world. The display and analysis of the volume of water in between the seafloor and the surface is one of the primary goals many marine GIS users are working to achieve.

With all of the modern technology available today, we are still limited in the capacity to map the undersea world, where laws of gravity are replaced by rules of density, temperature, salinity, and the currents that they influence. As the focus sharpens on the importance of marine resources, so do the tools that marine GIS users employ to represent the 71 percent of the surface of the earth that is water. In so doing, the marine GIS community will push the common perception of our oceans and seas beyond this surface, into the depths below.

People all over the world are using new technology to gather data above, as well as on and below, the sea surface: Echo-sounding instruments are used to bounce sound waves off the seafloor, providing a dense set of x,y, z-value points that can be used to create a bathymetric representation of the seafloor. Vessels check currents with special buoys to learn about patterns in current and temperature regimes. Sidescan sonar vehicles measure the backscatter of sound and can relay the data back to an onboard computer in real time. Researchers use satellite-tracking systems to follow marine species to learn of their migration patterns. Sub-bottom profilers take cross-section samples of the seafloor to see how deep the sediment layer extends, and what lies beneath it. Ecologists map the locations of coral reef ecosystems to track the influence of human activity on them. And many coastal applications benefit from light detection and ranging (LIDAR) technology that transmits and receives electromagnetic radiation at a high frequency to provide data with an absolute elevation accuracy of about 15 centimeters. The resulting data gathered for these applications demands a new and organized storage and analysis capacity, one that will encourage the development of tools to better understand and integrate the dynamic dimensions of the ocean environment.

In the pages that follow, *Marine Geography: GIS for the Oceans and Seas* will take you from the fringes of the world's coasts down to the bathymetric bottom. The resulting studies are organized into three sections that have been divided geographically by maritime region: the Atlantic Ocean, Pacific Ocean, and Northeast Pacific. Along the coasts of these bodies of water are some of the most heavily populated regions on earth. Population growth encroaching on coastal areas affects the marine ecosystem in ways that need to be studied, quantified, and presented to influence change. Maps, and more specifically digital maps, are fast becoming the most useful tool for this representation. The individual studies of this book demonstrate useful methodology, application, and insight that reveal the progress and innovative direction that people of this discipline are taking.

Recent initiatives in the discipline include the previously mentioned design of the marine GIS data models. In "The Inception of the ArcGIS Marine Data Model," we discuss this collaborative development effort that addresses the multiple dimensionality of oceanographic data, and new ways to assess the temporal and dynamic properties of shoreline and coastal processes. These dynamic boundaries can be modeled using spatial data structures that vary their relative positions and values over time and depth. The following works are presented according to their marine geographical region and successive order of appearance in the book.

Atlantic

Applying GIS to benthic marine systems and corals in the Caribbean, Eric Treml and his colleagues show that the spatial and temporal patterns in ecosystem dynamics are key to the effective conservation and management of reef systems. Their goal of improved conservation remains the focus for current and future research initiatives (including this case study of patch reefs around St. John) that may lead to future exploration of spatial patterns in coral community structure between islands in the eastern Caribbean and other island groups.

Michelle Kinzel and the Oceanic Resource Foundation's study of satellite telemetry and GIS literally rides on the back of two turtles to reveal the threats to their habitat and the turtles' migration routes around the perimeter of the Gulf of Mexico. Sea turtles face many threats while in their normal pattern of movement, and the integration of GIS analysis helps to reveal these factors and point the way toward effective conservation measures.

Richard Bates and Ben James, of the University of St. Andrews, examine a detailed and broadscale analysis of seafloor conditions within a marine Special Area of Conservation (SAC) in the waters of the United Kingdom. Data about the volume of water and the water quality within a marine SAC provides important information about the health of the sites involved.

Erik Franklin and his colleagues discuss a project for the Marine and Atmospheric Science Department at the University of Florida. GIS provides a powerful tool for their planning, implementation, and analysis of marine ecological projects and fisheries assessments. Delving into a variety of examples, they illustrate the diverse and useful ways in which a GIS can be integrated into a marine research and fisheries assessment program.

James McClean of the Florida Geological Survey investigates thermal data to identify underwater springs. The investigators are able to better plan and organize field operations by narrowing the potential survey areas. Using remote sensing imagery and conventional cartography data products, they are able to plan survey patterns and calculate time and fuel budgets in advance. This pilot project demonstrates how GIS can help to reduce research expenses and maximize efficient deployment of field survey teams.

Laura Engleby of the Dolphin Ecology Project provides some of the first research dedicated to bottlenose dolphins in Florida Bay. This provides a rich baseline of information for assessing the response of bottlenose dolphins to future changes in the south Florida ecosystem.

Continuing work documented of an ongoing study, Robert Schick reveals that both right whale and bluefin tuna movement exhibit significant correlation with sea surface temperature gradients. The response of each species is highly dynamic, pointing to future investigation of prey distribution and ocean characteristics to help explain how these species use their environment.

Implementing an enterprise geographic information system, Stuart Kininmonth at the Australian Institute of Marine Science describes his research methodology. With marine operations based on two ships, this system stores valuable data as the ships carry out scientific exploration of the marine environment during all seasons of the year.

Roger Goldsmith and A. J. Plueddemann of the Woods Hole Oceanographic Institution of Massachusetts describe how their GIS is being used to plan and refine the positioning of mooring locations. This study of Researcher Ridge, in a remote location northeast of South America, at a depth of 5,000 meters, was able to identify several alternative sites that might not have been so readily apparent without GIS analysis.

Doug Wilder and Henry Norris of the U.S. National Park Service take a close look at benthic environments. The Southeast Area Monitoring and Assessment Program–South Atlantic (SEAMAP–SA) established the Bottom-Mapping Workgroup in 1985 to gather, archive, and disseminate hard-bottom habitat data needed by researchers and managers to study and protect fish habitats.

Matthias Mueller and his colleagues with GEOMAR in Germany developed a Terrestrial and Marine Information System (TerraMarIS), a tool that helps to promote interdisciplinary research of the seafloor and investigate the coastal zones and the interior of the adjacent mainland.

Martin Kaye of the Bay of Fundy Resource Centre reports on the variety of opportunities offered by that organization's GIS support team. The center provides the facilities for meeting the challenges of the Bay of Fundy and the Gulf of Maine by designing viable mapping solutions for a large host of users.

Jack Sobel and The Ocean Conservancy use GIS to compile and utilize databases on marine and coastal protected areas (MACPAs) throughout the United States. Their initial GIS efforts were directed toward developing the first complete database of federal MACPAs and eventually were expanded to include state and regional information.

Pacific

Peter Rubec and Joyce Palacol of the International Marine Life Alliance investigate coral reefs and other coastal habitats throughout Southeast Asia. This ongoing study is of critical importance because invertebrates are being destroyed by destructive fishing methods. This work points to the use of GIS to assist with zoning coastal nearshore areas in a manner similar to terrestrial coastal planning.

Mark Abramson, Damon Wing, and their colleagues at Heal the Bay show how a small group of citizens can indeed fight city hall and make a huge difference in conservation policy. In their report, residents of the Malibu Creek Watershed area explain how they restored their local watershed and took action that resulted in statewide changes. Now, in its second decade, Heal the Bay continues its fight to find workable solutions to the problems threatening the future of the bay and all of Southern California's coastal waters.

Chad Nelsen and Mark Rauscher of the Surfrider Foundation describe how GIS and community activism played roles in developing Beachscape. This Surfrider Foundation program improves the availability of data along the coast and strengthens coastal management potential. Progressive mapping initiatives involving partnering organizations will help to create data standards to promote the sharing of spatial data for coastal areas.

Robert Burne and Christian Parvey of The Australian National University describe the development of a GIS-based project for the rapid appraisal of the benthic habitat domains of the Australian Ocean Territory (AOT). Four environmental models were developed from available data sets to provide a combined analysis leading to a variety of insights into habitat distribution.

Lance Morgan and Peter Etnoyer at the Marine Conservation Biology Institute of Redmond, Washington, provide a follow-up to their initiative presented in the Marine Geography section of *Conservation Geography.* Their goal is to generate a user-friendly GIS with interpretations for lay audiences as well as the scientific community.

In Australia, several marine parks and marine protected areas have come under the umbrella of the Great Barrier Marine Reef Marine Park—the largest marine park in the world. Trevor Gilbert, Tracey Baxter, and their colleagues at the Australian Maritime Safety Authority use environmental data, including a complete set of scanned and GIS-ready nautical charts for offshore sites and remote Australian islands and territories, in the course of their work.

Northeast Pacific

Creating a "conservation blueprint," Zach Ferdaña of The Nature Conservancy of Washington describes the "Ecoregional Planning Process." This study helps to define and illustrate a few methods for incorporating marine GIS into ecoregional planning, and spatially delineates a set of representative habitat and species assemblages that can be used for a variety of conservation efforts.

Bob Christensen gives a personal perspective on whales and other endangered marine life at Point Adolphus, Alaska. The Southeast Alaska Wilderness Exploration, Analysis and Discovery (SEAWEAD) group raises awareness for stewardship in local communities, conservation groups, management agencies, and businesses in the region.

Jeff Ardron of Living Oceans Society provides conceptual and technical direction in his analysis of physical benthic complexity. This is useful as an indicator of distinctive heterogeneous habitats often associated with species richness. Physical complexity is used to direct marine planning and research activities. The chapter concludes with a "recipe" for calculating benthic complexity based on bathymetric data.

Karen Dearlove and David Pray of the Conservation GIS Center in Anchorage, Alaska, report on the Alaska Oceans Network. The network is a voluntary association of conservation groups, environmental groups, fishing associations, and Alaska Native organizations that share a common goal to restore and maintain healthy marine ecosystems in Alaska. Its maps serve members, partners, and officials at state and federal levels and help illustrate environmental issues that affect Alaska's marine systems.

At the USGS Glacier Bay Field Station in Alaska, Philip Hooge describes how GIS is being used around glaciers. The field station provides marine GIS solutions to the public that were originally designed for Glacier Bay research projects. The designers had the foresight to make them adaptable by others in the marine GIS community. The ArcView® Animal Movement and Oceanographic Analyst extensions contain numerous functions designed to aid in marine analysis.

Philip Bloch and his colleagues at the People for Puget Sound, in Seattle, write about implementing marine science and policy. This chapter focuses on raising the awareness of resource users, managers, and the general public of the need to conserve, restore, and protect marine resources in the Orca Pass and other places that should be set aside as MPAs off the coast of Washington State.

The complex nature of the oceans and seas presents unique challenges to marine GIS users, and the chapters here describe how those challenges are being met with a host of exciting new mapping technologies. Potential future shifts in coastlines, due to global warming and consequent sea-level rise, challenge us to develop the methods and analytical tools to track this change and create sustainable solutions. *Marine Geography: GIS for the Oceans and Seas* brings together experts in this discipline, and serves as a tribute to the people working in the coastal zone, on the surface, in the water column, and on the seafloors of the world.

Marine Geography
GIS for the Oceans and Seas

Global Marine GIS

The Inception of the ArcGIS Marine Data Model

Joe Breman
ESRI
Redlands, California

Dawn Wright
Department of Geosciences, Oregon State University
Corvallis, Oregon

Patrick N. Halpin
Nicholas School of the Environment and Earth Sciences, Duke University
Durham, North Carolina

The increased commercial, academic, and political interest in coastal regions, oceans, and marginal seas has spurred fundamental improvements in the analytical potential of GIS, while extending the methodological framework for marine applications. Many challenges remain, such as addressing the multiple dimensionality and dynamism of oceanographic data, handling the temporal and dynamic properties of shoreline and coastal processes, dealing with the inherent fuzziness of boundaries in the ocean, the great need for spatial data structures that vary their relative positions and values over time, and, last but not least, the development of effective conceptual and data models of marine objects and phenomena.

Increased interest in marine and coastal applications of GIS led to the first-ever ESRI marine special interest group (SIG), which met at the ESRI® International User Conference in San Diego in July 2001. Members of the marine community expressed the need for a marine data model to facilitate the representation and analysis, in digital form, of features along coasts and in estuaries, marginal seas, and the deep ocean. A few months later, in the fall of 2001, leaders of the discipline from professional, academic, and government organizations came together for a workshop at ESRI in Redlands to contribute various perspectives toward the creation of a marine data model.

Why a marine data model?

As noted by Bartlett (2000), one of the most important lessons to be learned from collective experience in the application domain of marine GIS is capable planning and designing of the data model before attempting to implement a GIS database. With regard to ESRI products, many marine and coastal practitioners and organizations use the coverage data model. Although this has largely been successful, coverages have limitations with respect to topology. For example, features are aggregated into homogenous collections of points, lines, and polygons with generic, one- and two-dimensional "behavior" (Zeiler 1999). It has become more vital to distinguish between behavior within feature classes. For example, the behavior of a point representing a marker buoy is identical to that of a point representing a pulsing transponder; the behavior of a line representing a road is identical to the behavior of a line representing a dynamic shoreline. One of the goals of the marine data model is to provide a structured framework that more accurately represents the dynamic nature of water processes. Shapefiles and coverages can now be easily loaded as feature classes in the ArcGIS™ geodatabase for more organized, scalable, and flexible analysis. The marine data model development will provide data managers with a ready-made model for marine data so that they can spend more time at sea collecting data and in the lab analyzing their information, and less time at the computer planning the data structure. It should also be useful to GIS practitioners working as academic, government, or military oceanographers, coastal and marine resource managers, consultants, technologists, archaeologists, conservationists, geographers, fisheries managers, scientists, ocean explorers, and mariners alike.

ArcGIS 8 introduces the geodatabase, an object-oriented data model in which GIS features are smarter; they can be endowed with real-world behaviors for individual objects within the same categories, and relationships, domains, and ranges may be defined among them. (For an overview of ArcGIS object and geodatabase concepts see Zeiler 1999). This has exciting implications for marine and coastal applications, but questions and concerns in the community have surfaced. Given the existing investment, how and when should one make the transition to object-oriented ArcGIS 8? How well are marine application domain requirements met in the geodatabase structure now? What can the community do as a group to understand the technology and identify requirements? What are the potential benefits?

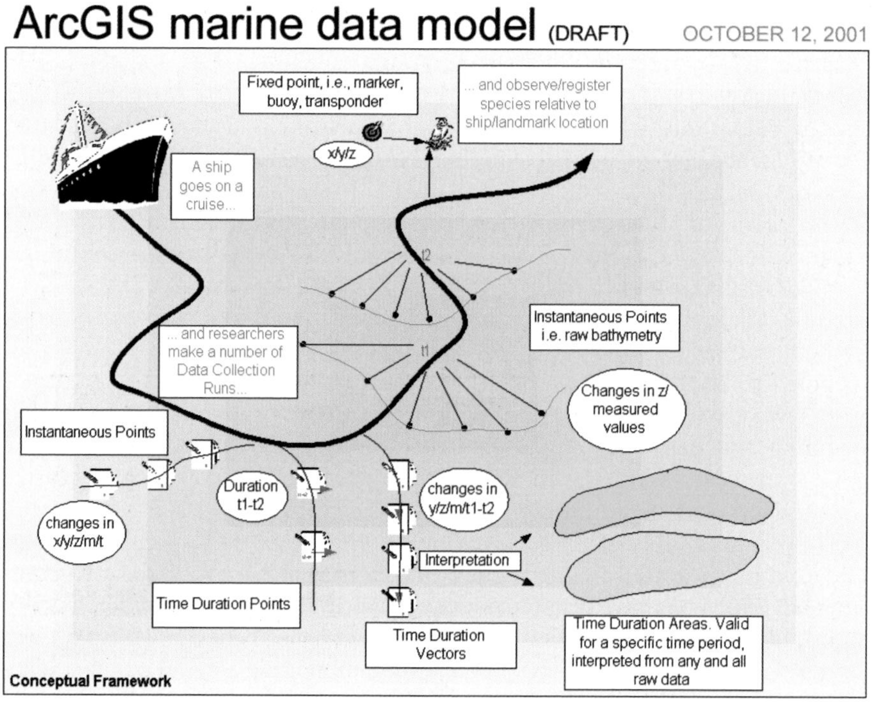

Figure 1: A schematic draft of the conceptual framework portion of the ArcGIS marine data model created by ESRI developer Steve Grisé. This diagram gives a visual representation of the elements of the data model as they relate to important aspects of marine data collection and analysis.

The ArcGIS Data Model working group was initiated to help address these concerns and to support the marine community in making this important transition. The progress of the working group, as well as the general development of the model, and related Microsoft® PowerPoint® slide presentations, can be found on the Web at *dusk.geo.orst.edu/djl/arcgis.* The site includes a form visitors may use to record their interest in the marine data model. Of the initial hundred visitors to the site, a group of 20 was selected to form a review panel for the model. And in addition to development of the data model itself, a comprehensive draft conceptual framework document (upon which this chapter is largely based) was also prepared (Wright et al. 2001), providing a descriptive analysis of the data model development and evolution.

Specific features of the model

A key advantage of the ArcGIS marine data model is that it will help users employ the more advanced manipulation and analysis capabilities of ArcGIS, particularly the ability to capture the behavior of real-world objects in a geodatabase, and the support of more complex rules that can be built into the geodatabase. For users, the data model provides a basic template for implementing GIS projects (i.e., inputting, formatting, geoprocessing, and sharing data; creating maps; performing analyses; etc.); for developers, it provides a basic framework for writing program code and maintaining applications. And while

Figure 2: The common marine data types listed here are the result of a brainstorming session at the first marine data model workshop. This list is evolving and changing with review of the data model.

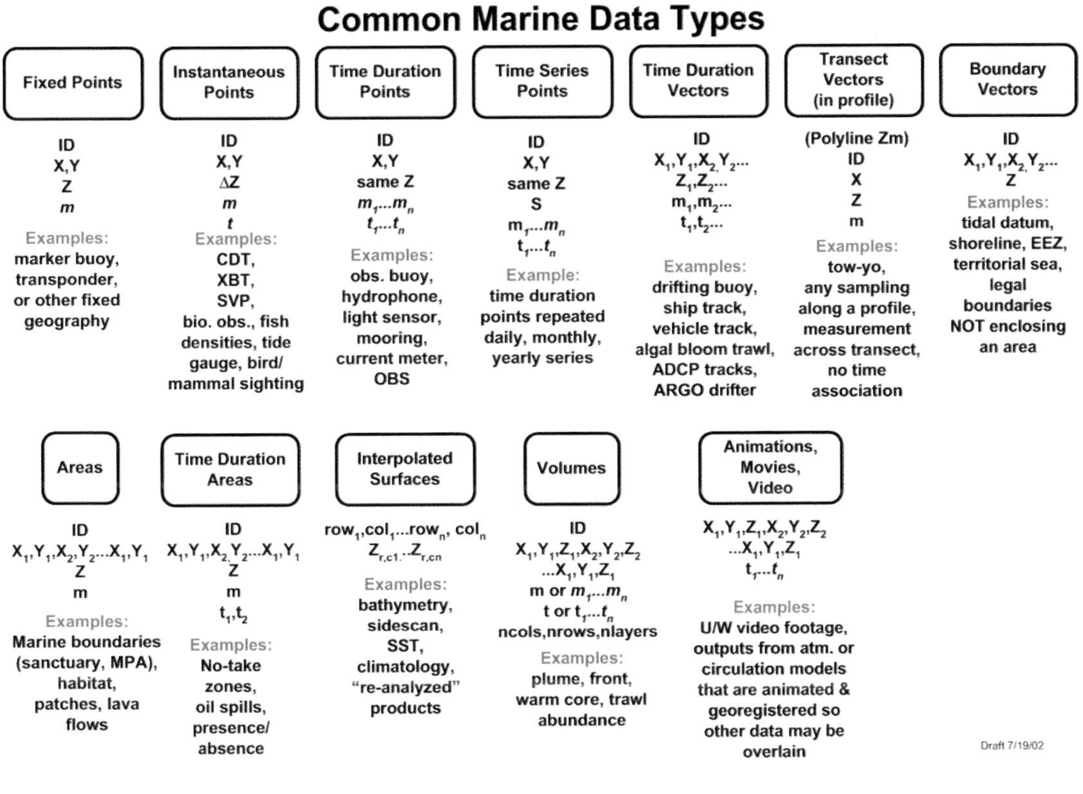

ArcGIS data models do not create formal data standards, they do promote existing ones, to simplify the integration of data at various jurisdictional levels (i.e., local, state/provincial, national, global). Specific goals include:

- Production of a common structure, a "geodatabase template," for assembling, managing, and publishing marine data in ArcGIS.
  1. For example, the model is specified in an industry-standard modeling notation called the Unified Modeling Language (UML). And because UML code is easily converted to an ArcGIS geodatabase, users can immediately begin populating the geodatabase rather than having to design it from scratch.
  2. Users can produce, share, and exchange data in similar formats.
  3. Unified approaches encourage development teams to extend and improve ArcGIS software.
- Extending the power of marine GIS analyses by providing a framework for incorporating behaviors in data, and dealing more effectively with scale dependencies.
- Providing a mechanism for the implementation of data content standards, such as the Federal Geographic Data Committee (FGDC) Hydrography Data Content Standard for Inland and Coastal Waterways, critical for the Coastal National Spatial Data Infrastructure.

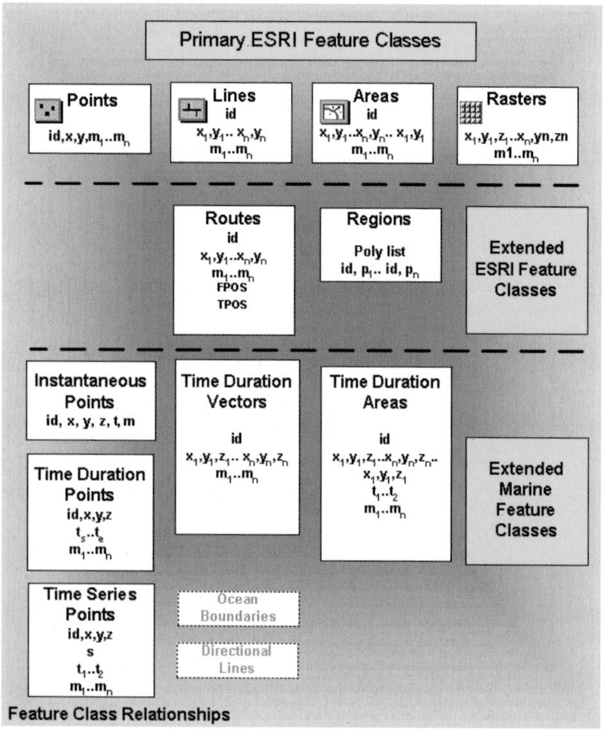

Figure 3: The primary ESRI feature classes are the building blocks of the data model. These are the most commonly used features that provide a topological structure for the underlying data.

In summary, the marine data model will also help marine GIS users to learn and understand ArcGIS. For instance, in the core ArcGIS data model, arc–node topology can be used in conjunction with other new and powerful data structures. Routes and regions are relegated to feature classes. And relationships between tables can be preserved, maintained, and easily managed. Most importantly, already existing shapefile and coverage data can be loaded into the geodatabase for more efficient and effective data management by the marine community as a whole.

The marine data model represents a new approach to spatial modeling in a way that promotes better integration of many important features of the ocean realm, both natural and artificial, and provides more accurate representations of location and spatial extent, along with a means for conducting more complex spatial analyses of the data. The model also considers how marine and coastal data might be more effectively integrated in space and time, particularly into the all-important third and fourth dimensions. The data model, although still limited to 2.5-D, includes "placeholders" to represent the fluidity of ocean data and processes.

Because many marine activities require depth values (z) as an integral part of data collection and analysis, the marine data model attempts to explicitly define z-values in the geometry of the features. Point features typically have an x and y field, and most often in the case of marine data a "z-value" field must be appended for depth. The marine data model includes "z-aware" features; these are points that recognize the z-value as a part of their "shape" field.

As measurements gathered at sea are usually time dependent, different types of time-dependent features are included in the data model. The suggested data model structure explicitly distinguishes between geographic features associated with measurements at a specific point in time (e.g., instantaneous points) versus features that are associated with attributes that are collected over a duration of time (e.g., time duration points, vectors, or areas). Data collected on the coast

Figure 4: With improved technology comes the ability to collect data in many new ways. This visual representation features many of the data collection methods available today. At the Coastal Geo-Tools Conference in Charleston, South Carolina, in 2001, Jack Dangermond, president of ESRI, spoke of floats that may be employed in the future to gather data from the ocean surface to the seafloor.

IMAGE COURTESY OF COASTAL OCEAN OBSERVATION LAB, INSTITUTE OF MARINE AND COASTAL SCIENCES, RUTGERS UNIVERSITY, NEW BRUNSWICK, N.J. MARINE.RUTGERS.EDU/COOL

and shorelines, for example, often must attribute tidal variance, wave activity, and even atmospheric pressure; all can be more effectively incorporated into the marine data model using time duration points. This principle can also be applied to lines. A trawl or observation run has a beginning and end time used to calculate the effort expended in the collection. A "time duration vector" *(figure 3)* accommodates this type of marine feature class. A primary goal of the marine data model is to make it easier to create spatial data sets that reflect specific differences in data collection and analysis needs of the marine GIS user community not currently addressed by standard GIS models.

Conclusion

Although the focus of the ArcGIS marine data model is on both the deep ocean and the coast, it attempts to represent the essential elements for a broad range of marine and coastal data types and processes. One model cannot include a comprehensive catalog of objects meeting the needs of all user groups and applications. This model is a starting point upon which to build and leverage the experiences of a broader range of practitioners—much broader, in fact, than the specialties of the initial Marine Data Model working group. It will also be desirable to understand the similarities that may exist between the marine model and related data modeling efforts in hydrology, biodiversity and conservation, and land parcel management. Steve Grisé, ESRI development lead for the marine and other data models, says that designing a data model is like designing a new minivan: once it's designed and built you still don't know how a specific family will actually use it. Once a data model is designed and initially implemented, then its most useful applications will become clearer, and it can be refined accordingly.

Acknowledgments

The authors would like to acknowledge the core marine data model team, as well as the larger review team, for their active participation and ideas, which continue to be essential in the development of this project. We would also like to encourage marine and coastal specialists, as well as other GIS professionals, to engage in discussions on the use and application of this model, especially through the online resources of the model's Web site: *dusk.geo.orst.edu/djl/arcgis.*

References

Bartlett, D. J. 1993a. Space, time, chaos, and coastal GIS. *Proceedings of the International Cartographic Conference,* Cologne, Germany, 539–51.

Bartlett, D. J. 1993b. Coastal zone applications of GIS: Overview. In *Explorations in Geographic Information Systems Technology Volume 3: Applications in Coastal Zone Research and Management,* edited by K. St. Martin. Worcester, Mass.: Clark Labs for Cartographic Information Systems Technology and Analysis, and Geneva, Switzerland: United Nations Institute for Training and Research (UNITAR), 1–18.

Bartlett, D. J. 2000. Working on the frontiers of science: Applying GIS to the coastal zone. In *Marine and Coastal Geographical Information Systems,* edited by D. J. Wright and D. J. Bartlett, 11–24. London: Taylor & Francis.

Forestry Special Interest Group (FSIG). 2000. Forestry data model. ESRI, Redlands, Calif. Retrieved from *arconline.esri.com/arconline/datamodels_one.cfm?id=10*

Li, R. 2000. Data models for marine and coastal geographic information systems. In *Marine and Coastal Geographical Information Systems,* edited by D. J. Wright and D. J. Bartlett, 25–36. London: Taylor & Francis.

Li, R., and N. K. Saxena. 1993. Development of an integrated marine geographic information system. *Marine Geodesy* 16:293–307.

Lockwood, M., and R. Li. 1995. Marine geographic information systems: What sets them apart? *Marine Geodesy* 18(3):157–59.

Wright, D. J., and D. J. Bartlett, eds. 2000. *Marine and Coastal Geographical Information Systems.* London: Taylor & Francis.

Wright, D. J., and M. F. Goodchild. 1997. Data from the deep: Implications for the GIS community. *International Journal of Geographical Information Science* 11(6):523–28.

Wright, D. J., P. N. Halpin, S. Grisé, and J. Breman. 2001. ArcGIS marine data model, ESRI, Redlands, Calif. Retrieved from *dusk.geo.orst.edu/djl/arcgis*

Zeiler, M. 1999. *Modeling our world: The ESRI guide to geodatabase design.* Redlands, Calif.: ESRI Press.

Atlantic

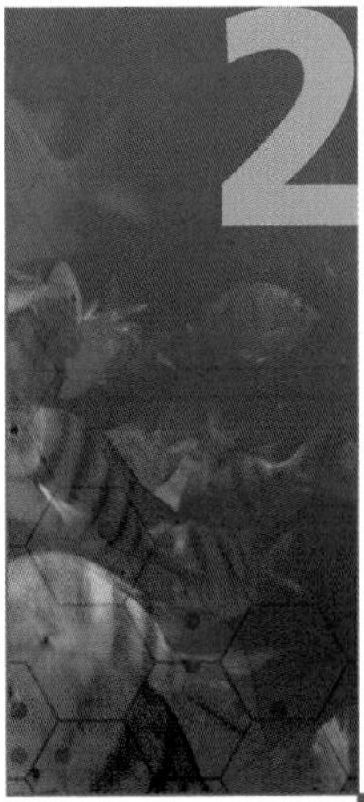

Spatial Ecology of Coral Reefs

Applying Geographic Information Science to Benthic Marine Systems

NICHOLAS SCHOOL OF THE ENVIRONMENT AND EARTH SCIENCES
DUKE UNIVERSITY

Eric A. Treml and Patrick N. Halpin
Nicholas School of the Environment and Earth Sciences, Duke University
Durham, North Carolina

Mitchell Colgan
University of Charleston
Charleston, South Carolina

The world's coral reefs are under continual stress by the accumulation of natural and anthropogenic disturbances. Fortunately, there may be new hope as the importance of these valuable ecosystems is being realized within the U.S. federal and international political arenas. With new urgency, many institutions are setting out with heightened attention to protect and monitor these systems. To effectively manage and protect coral reef communities, researchers need to quantify the hierarchical linkages and cumulative effects of environmental stressors on reef systems, along with the ecological connectivity between individual reef sites. Research that addresses the dynamics of coral reefs in time and space may provide a better methodology for studying these complex ecosystems, ultimately leading to better management. This chapter presents a spatial approach to coral reef science and describes a case study from the U.S. Virgin Islands: an investigation of the linkages between environmental stressors (sedimentation, wave energy, and hurricanes) and their scale-dependent impact on the fringing coral reef communities.

Introduction

Coral reefs are estimated to cover more than 600,000 square kilometers (Smith 1978), along 20 percent of coastlines globally in more than one hundred countries. These dynamic and incredibly diverse ecosystems help build islands, expand shorelines (Birkeland 1997), sustain economies, and provide sustenance for many tropical nations. Unfortunately, 11 percent of the world's reefs have been lost, 15 percent are critically endangered (Wilkinson 2000), and nearly 60 percent are threatened by human

Example of the invertebrate diversity found on coral reefs. Photo taken at 3 meters depth—Komodo, Indonesia.

PHOTO COURTESY OF ERIC TREML

PHOTO COURTESY OF ERIC TREML

Coral outcropping, taken at 10 meters depth—Komodo, Indonesia.

activities, according to the U.S. Coral Reef Task Force (USCRTF 2000). The global community has recognized the need to effectively manage and conserve coral reefs, and a change in political climate now favors research directed at increasing the understanding of how disturbances, acting at different spatial and temporal scales, affect coral reefs. Priorities for coral reef research and the needs of the coastal management community often include the following (Crosby et al. 2000, Wilkinson 2000, Salm et al. 2000, McManus 2001):

- Research should be cross-institutional, interdisciplinary, and fully integrated with policy and decision making.
- Monitoring and research programs should be expanded in scale and scope.
- Monitoring and research should be conducted as a multiscaled and hierarchical process.
- There is a need for increased local involvement in monitoring and management.
- The use of marine protected areas (MPAs) should be expanded, with increased coordination and linking of individual sites into networks.
- To completely protect fish and coral communities, the use of multiuse zones within MPAs should be explored.
- Traditional rights and management practices need to be recognized.

Central to many of these requests is the need for spatially explicit research on the impact of environmental stressors, both natural and anthropogenic, and the role of the physical and biological environment on reef systems. In addition, there is the need for understanding ecological connectivity between reefs and their coastal watersheds, between neighboring (upstream/downstream) reefs, and between individual reefs and their regional setting.

The general priorities of spatial ecology research on coral reefs should include many elements. The accurate mapping of reef and related systems is essential. The identification of the hierarchical scaling and effects of environmental stressors on coral reef community structure is assisted by qualitative investigation. By quantifying the spatial and temporal ecological connectivity between reef systems, including the degree to which a reef system is open or closed to exogenous factors (larval immigration, sediments, and so forth), we can investigate the influence of the spatial arrangement. The size of the marine protected areas, or other spatially and temporally explicit management options, must take into consideration ecological connectivity and scaling of environmental stressors. Finally, it is critical to bring the implications from this spatial research to the MPA community as tools for use in its decision making.

Success of this science-based coral reef protection initiative depends on a solid understanding of reef science. The initiative needs to address and include life-history characteristics, coral reef community ecology, and the interrelationships between the target organisms and the physical marine environment, including the influence of natural and anthropogenic disturbances.

Research needs

The corals responsible for the development of the physical reef structure are stationary and light dependent; as a result, they possess similarities with terrestrial plant communities. For this reason, it has been commonplace to adopt concepts and techniques used by plant ecologists for the study of coral reef communities (Loya 1972, Connell 1978, Sousa 1985). Integrating techniques found in terrestrial ecology with advanced spatial technologies (i.e., remote sensing, GIS, and modeling) in the study of coral reef ecology and management may provide a better methodology for addressing the complex biological and physical dynamics defining present-day coral reef ecosystems. This landscape ecological approach addresses multiscale mechanisms, including disturbances, and how they affect communities (Urban et al. 1987, Wiens 1989, Turner et al. 1991).

Young Elkhorn coral—St. John, U.S. Virgin Islands.

PHOTO COURTESY OF ERIC TREML

Many studies have emphasized the importance of disturbance in structuring and maintaining coral reef communities (e.g., Adey and Burke 1977, Hubbard 1986, Brown 1997). Decades of studies reveal a paradigm: coral reefs are stable and quite persistent on a geologic time scale, yet the community structure is variable and highly irregular on an ecological time frame (over a period of decades). Reef building has occurred for nearly two hundred million years, withstanding sea level fluctuations, numerous periods of glaciation, and several mass extinction events. However, ecological studies show that reefs are disturbed systems, vulnerable to stresses, and may be in decline (Jackson 1997, Wilkinson 2000). A better understanding of how these disturbances affect coral reefs is greatly needed (Jackson 1991, Hughes 1994).

The reef community's location, shape, and very existence result from dynamic processes (such as hurricanes, sedimentation, wave energy, light, recruitment, dispersal, and competition) integrated through space and over time (Hubbard 1997). Few studies of coral reef systems have explicitly addressed this scale-dependent nature of processes and the resultant patterns (but see Murdoch and Aronson 1999). The effectiveness of cross-scale studies depends on whether the process of interest changes noticeably within the scale of the study. Many effective studies of coral reefs addressing ecosystem process and patterns have been completed, although within narrow spatial or temporal scales or both. These studies form the solid foundation needed for the synthesis and future research into space–time scaling issues of ecological processes (Hatcher 1997).

The challenge associated with the interrelationship between pattern, process, and scale is a central problem in ecology, often unifying basic and applied ecology (Hughes 1994, Levin 1992, Hutchinson 1953). To understand the way communities are organized and how they function, researchers must strive to find patterns that can be quantified. Pattern analyses, first used by terrestrial ecologists (Watt 1947, MacArthur and Levins 1964), are techniques used to detect and quantify nonrandom distributions of organisms. Once patterns are described, researchers can seek to identify the mechanisms that generate and maintain those patterns (Hatcher et al. 1987, Turner 1989, Gustafson 1998), although, statistically, describing patterns and demonstrating that a specific mechanism can give rise to the observed patterns does not prove that the proposed mechanism is definitely responsible. Furthermore, no single mechanism explains patterns on all scales (Wiens 1989); different processes may be important at different or multiple scales. Therefore, the concepts of pattern and scale are inseparable (Hutchinson 1953, Hatcher 1997). For example, sedimentation from coastal watersheds may explain the patterns in coral community structure at the scale of bays (hundreds of meters) but fails to explain the patterns at the scale of an archipelago (thousands of meters); here, wave energy or storm frequency may best explain the patterns in community structure. Using the fundamentals of landscape ecology with the capabilities of GIS, remote sensing, and spatial modeling, the potential exists to greatly improve coral reef research efforts.

Case study: St. John, U.S. Virgin Islands

In an attempt to better understand the complex interrelationship between physical forces and the spatial patterns present in coral communities at the local (island-level) scale, 14 discrete patch-reef communities were studied along the south shore of St. John, U.S. Virgin Islands. This study quantified variation in coral community structure at multiple spatial scales: among transects within patch reefs (10–100 meter scale), among patch reefs within bays (100–500 meter scale), and between bays (500–1,000 meter scale). Specifically, this research addressed the following questions: What physical force(s) explains the variation present in coral community structure within the area? Does the terrestrial coastal zone (watershed characteristics and potential runoff) have a primary influence on the coral reef community structure? Or, does the hydrographic regime (dominant currents, exposure, hurricanes, or some combination thereof) have a greater influence on community structure?

PHOTO COURTESY OF ERIC TREML

Europa Bay. St. John's steep terrain and rugged shoreline are apparent. Watersheds along the south shore and within the national park boundaries are predominately undeveloped.

Study site

The U.S. Virgin Islands are located in the northeastern Caribbean, approximately 80 kilometers east of Puerto Rico. The island of St. John *(figure 1)* has an area of approximately 48 square kilometers, consisting of very steep terrain and a rugged coastline. The Antilles Current, which flows from the southeast at approximately 10–20 centimeters per second (NOAA drifter buoy data), dominates the current regime for the northeastern Caribbean. The wave energy is concentrated around projecting headlands and within east- and south-facing bays. The study site for this research is located along the south shore of St. John. The entire site is contained within the boundaries of the Virgin Islands National Park and Biosphere Reserve. The total marine environment covered in this study is approximately 4 square kilometers and includes a variety of benthic environments, such as seagrass meadows, sand channels, coral reefs, algal reefs, and rubble zones.

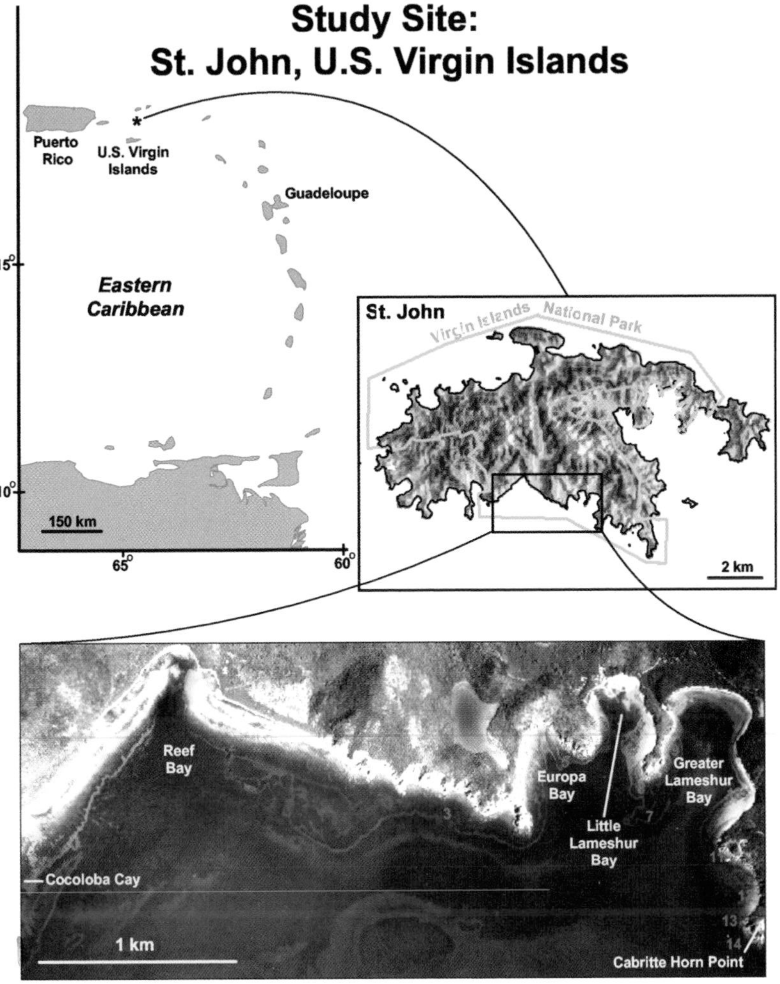

Figure 1: The study area is located along the south shore of St. John, U.S. Virgin Islands, just east of Puerto Rico. The study site is bound by Cabritte Horn Point and Cocoloba Cay. Individual patch reefs are numbered: (1) W. Reef; (2) E. Reef; (3) White Cliffs; (4) W. Europa; (5) E. Europa; (6) E. Lt. Lameshur; (7) Yawzi Carb; (8) Yawzi Rock; (9) W. Gr. Lameshur; (10) Donkey Bight; (11) Tektite Rock; (12) Beehive Cove; (13) Cabritte Horn Pt.; (14) Tektite Reef.

PHOTO COURTESY OF ERIC TREML

A modified line transect method was used to determine substrate characteristics and coral community structure.

Methods

Several techniques were used to analyze the marine benthic communities and the primary controls on reef development. Prior to arriving on St. John, all fringing patch reefs within the study area were located and digitized using scanned aerial photography, based on the distinct sand halos surrounding each reef. In the field, a modified line transect method (Loya 1972) was used to quantify reef community characteristics, record depth, and measure substrate characteristics. More than 3,000 meters of reef along 92 transects were surveyed and used to calculate community indexes: species richness (total number of species encountered), percent total coral cover, percent total algae cover, coral diversity, relative species abundance (percent cover of each species), and coral community evenness.

The long-term average sediment delivery on St. John was estimated using a simplified version of the revised universal soil loss equation (Wischmeier 1976, Anderson 1995, Anderson and MacDonald 1995). Due to the homogeneous forest cover and uniform soil conditions, watershed size and average slope can be used as an indicator of the relative, long-term average of natural sediment yield for the coastal watersheds (Hubbard et al. 1987, D. Anderson, 1998. E-mail.). Watershed size, slope, and drainage were all determined through the analysis of a high-resolution digital elevation model (DEM) using ArcInfo™ software from ESRI and its hydrologic modeling capabilities. The location of the stream outlets and the watershed characteristics (size and slope) were used as a proxy for long-term average sedimentation *(figure 2)*.

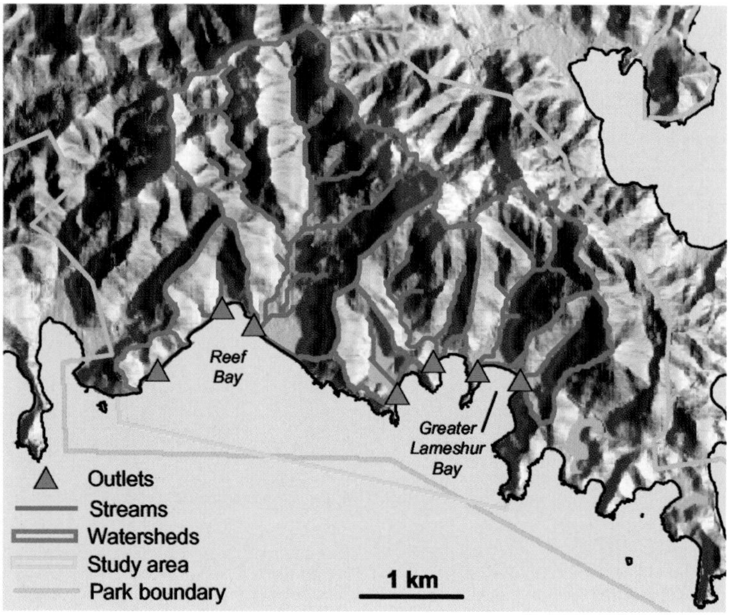

Figure 2: Hydrography of St. John's south shore. This map shows the major watersheds, stream locations, and outlets within the study site.

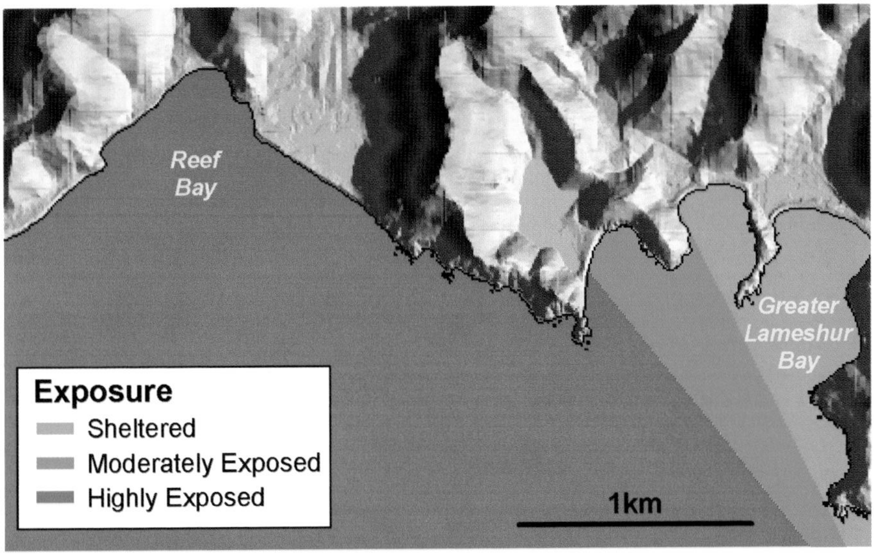

Figure 3: The study site was categorized into three exposure regimes, based on the velocity readings from current meters and flow directions from drogue deployments, NOAA Drifter Buoy composites, and personal observations in the field.

Average current directions and velocities have been acquired from the literature (Hubbard et al. 1987, Rogers and Teytaud 1988) and modern ocean current models for the Caribbean (NOAA drifter buoy data). Water velocities and local currents along the south side of St. John have been obtained from previous studies (Witman 1992), and unpublished current meter and drogue results from P. Edmunds (1998). The degree to which a patch reef is sheltered from these currents and wave energy was ranked relative to each other *(figure 3)*.

The single-classification analysis of variance (ANOVA) and the Tukey–Kramer multiple comparisons test were used to indicate significant difference in community indexes (percent coral cover, coral species richness, and coral diversity). The Bray–Curtis similarity index was used to statistically cluster the species-specific percent cover. For detailed methods see Treml (2000).

Metric	W Reef (1)	E Reef (2)	White Cliffs (3)	W Europa (4)	E Europa (5)	E Lt. Lameshur (6)	Yawzi Carb (7)	Yawzi Rock (8)	W. Gr. Lameshur (9)	Donkey Bight (10)	Tektite Rock (11)	Beehive Cove (12)	Cabritte Horn Pt. (13)	Tektite Reef (14)
Species richness	13	13	14	12	10	5	9	16	7	5	12	12	9	14
Percent coral cover	6.4	8.6	10.6	7.4	5.4	2.5	27.2	14.8	21.0	43.1	17.0	13.3	44.9	32.0
stdev+/–	3.3	5.2	4.2	3.3	4.8	1.2	11.1	3.8	6.8	37.0	6.3	10.9	7.3	13.0
Diversity	0.29	0.30	0.35	0.28	0.17	0.08	0.37	0.55	0.26	0.21	0.52	0.23	0.61	0.91
stdev+/–	0.14	0.13	0.12	0.14	0.11	0.02	0.12	0.10	0.05	0.09	0.15	0.11	0.16	0.27
Percent algae cover	24.1	27.1	11.2	35.1	2.6	59.9	53.6	25.9	11.7	8.9	28.7	2.2	36.4	27.5
stdev+/–	26.8	32.0	7.7	34.0	2.4	30.6	15.7	16.3	5.6	9.4	7.6	3.1	12.4	7.6

Table 1: Community Indexes. Quantitative coral community descriptions, based on the transect sampling method, are summarized by patch reef name.

Results

Transect data results, summarized at the patch level, are presented in table 1. The ANOVA and the Tukey–Kramer matrix of significant difference indicate that the 14 patch reefs consist of six statistically unique groups *(figure 4).* With the exception of the Greater Lameshur Group, this grouping strategy appears to be robust. In addition, the species-specific Bray–Curtis clustering analyses were consistent with the ANOVA results. The family of statistical trees produced by (a) removing the rare species, (b) removing the dominant species, and (c) including all species, were similar. The differences between these trees were in the placement of two patches within the Greater Lameshur Group (Donkey Bight and Beehive Cove), and the Cabritte Horn Point patch reef *(figure 5).*

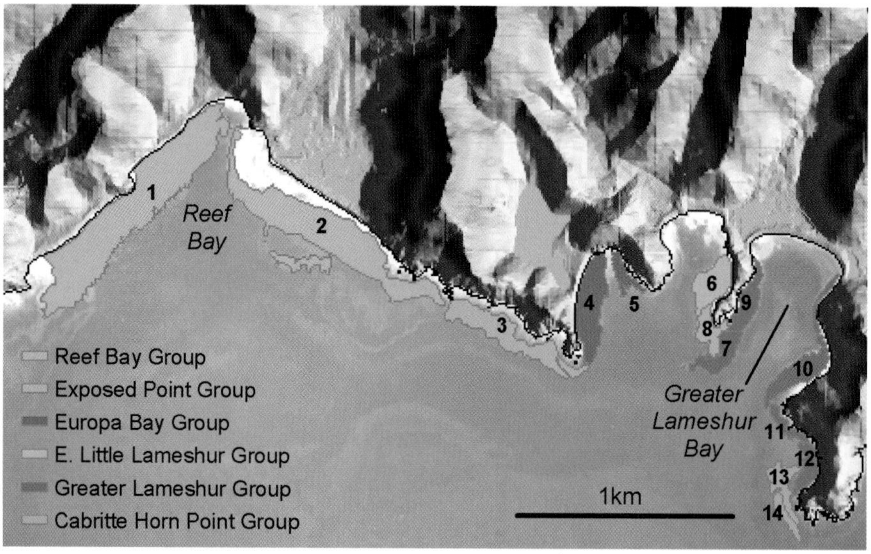

Figure 4: Map of patch reef groups based on ANOVA and Tukey–Kramer matrix. Patches within groups are not significantly different; between-group differences are significant.

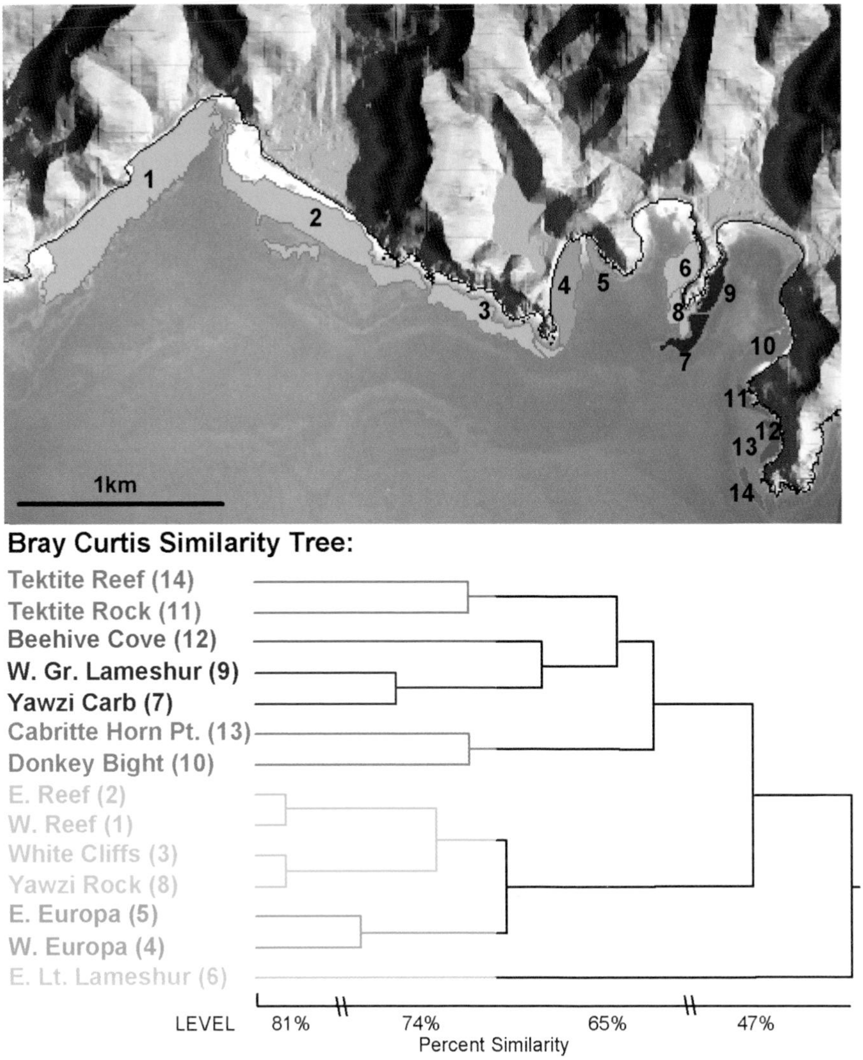

Figure 5: Bray–Curtis clustering results. Percent similarity based on species-specific percent cover. Similarity values calculated using the unweighted pair-group mean average combinational sorting strategy. Dendrogram is color-coded to reference the map of patch reefs within the study area.

Conclusion

The study took place seven years after Hurricane Hugo (category IV storm, 1989). The reefs of St. John were in an early phase of recovery following Hurricane Hugo's devastating blow to the reef corals; species diversity and total percent coral cover were reduced and dominant corals were nearly eliminated from the study area (Rogers et al. 1997). The research of coral community structure along the south side of St. John examined processes acting across multiple spatial scales but at a single point in time. Although this study was limited on the temporal scale, it documents the variability across a broader spatial scale than previous studies conducted on St. John. Integrating previous studies with this research enables a detailed broad-scale examination of the reef dynamics, which builds a clearer picture of the coral community dynamics.

The results of the study reveal the widespread lack of recovery from Hurricane Hugo along the south side of St. John. Although historical hurricane damage may account for the uniformly low-percent coral cover, it cannot account for the variability in coral community structure between individual patch reefs and between bays. On St. John, the primary physical forces acting at the spatial scale from bays to patch reefs include watershed-based runoff and subsequent sedimentation, and wave energy produced through a combination of ocean currents and degrees of wave exposure (Hubbard et al. 1987, Jackson 1991, Hubbard 1988).

The highest degree of variability was found between bays, equivalent to the kilometer-wide scale. Coral cover, coral diversity, species richness, and species-specific cover varied little at the transect level (within discrete patches), yet varied significantly from patch to patch. Generally, patch reefs within bays, and in close proximity, were more similar than patch reef communities compared between bays. This hierarchical pattern in variability can be shown as follows:

within patch <<	between patches / within bays <<	between bays
(low variability)		(great variability)

This pattern indicates that community structure may be driven by one or several controlling factors acting at spatial scales equal to the between-bay scale.

During the early stages of recovery, when space is not limited, physical factors such as wave energy (Adey and Burke 1977), bay geometry (Hubbard et al. 1987), orientation to dominant currents, and depth (Witman 1992), may control the development and maintenance of coral community structure. The species-specific percent cover cluster analysis *(figure 5)* shows this pattern clearly. The two primary branches (excluding East Little Lameshur) of the similarity tree correspond with the degree of exposure. The results of this case study show that the variation in the coral reef community structure along the south side of St. John is most likely due to the wave energy gradient produced by the dominant currents and the sheltering effect of projecting headlands.

Summary

Rarely will a single factor (e.g., hurricanes, currents, or sedimentation) explain all the patterns found in and between coral reef communities. More often, a suite of controlling factors (for any particular location) influences the community structure at multiple spatial and temporal scales. Therefore, the spatial and temporal scale of observations, as well as the scales at which physical forces influence a reef community, must be considered for the appropriate interpretation of ecological data. With a rising need and interest in multifactorial, broad-spatial, and long-term sampling and monitoring efforts, the research community may benefit by utilizing a more spatially explicit approach to coral reef ecology and management. Using this approach with the technologies of GIS, remote sensing, and spatial modeling, coastal managers and scientists may be better equipped to visualize and prepare for the complex changes taking place within coral reefs now and into the future (Hughes 1989, Jackson 1991).

Understanding the spatial and temporal patterns in ecosystem dynamics is essential to the effective conservation and management of reef systems. This goal of improved conservation remains the focus of our current and future research initiatives. To accomplish this we plan to expand this case study to all the patch reefs around St. John; explore spatial patterns in coral community structure between islands in the Eastern Caribbean and other island groups; quantify ecological connectivity between reef systems, including larval dispersal and retention; develop a hierarchical spatial data model for marine conservation; and investigate designs for a marine protected area network, based on the spatial ecology of coral reefs.

Acknowledgments

The authors thank Renée Seaman for her critical review and editorial comments in the development of this chapter. The case study was partially funded by a South Carolina Space Grant Fellowship, University of Charleston Deep Water Fellowship, and the University of the Virgin Islands, Virgin Island Environmental Resource Station.

References

Adey, W. H., and R. B. Burke. 1977. Holocene bioherm of Lesser Antilles: Geologic control of development. In *Reefs and related carbonates: Ecology and sedimentology,* edited by S. H. Frost, 67–81. Tulsa, Okla.: American Association of Petroleum Geologists.

Anderson, D. M. 1994. Guidelines for sediment control practices in the insular Caribbean. *Caribbean Environment Progress Technical Report* 32:63.

Anderson, D. M., and L. H. MacDonald. 1995. An investigation of sediment sources affecting marine resources at Virgin Islands National Park. *Park Science* 15(2):26–28.

Beets, J., L. Lewand, and E. S. Zullo. 1986. Marine community descriptions and maps of bays within the Virgin Islands National Park/Biosphere Reserve. *Biosphere Reserve Research Report* 2:118.

Birkeland, C. 1997. Introduction. In *Life and death of coral reefs,* edited by C. Birkeland, 1–12. New York: Chapman and Hall.

Brown, B. E. 1997. Disturbances to reefs in recent times. In *Life and death of coral reefs,* edited by C. Birkeland, 354–79. New York: Chapman and Hall.

Connell, J. H. 1978. Diversity in tropical rain forests and coral reefs. *Science* 199:1,302–9.

Crosby, M. P., K. S. Greenen, and R. Bohne. 2000. *Alternative access management strategies for marine and coastal protected areas: A reference manual for their development and assessment.* Washington, D.C.: U.S. Man and the Biosphere Program.

Edmunds, P. J., and J. D. Witman. 1991. Effect of Hurricane Hugo on the primary framework of a reef along the south shore of St. John, United States Virgin Islands. *Marine Ecological Progress Series* 78:201–4.

Edmunds, P. J. 1998. Letters and data printouts.

Gustafson, E. J. 1998. Quantifying landscape spatial pattern: What is the state of the art? *Ecosystems* 1:143–56.

Hatcher, B. G. 1997. Coral reef ecosystems: How much greater is the whole than the sum of the parts? *Coral Reefs* 16, Supplement: S77–91.

Hatcher, B. G., J. Imberger, and S. V. Smith. 1987. Scaling analysis of coral reef systems: An approach to problems of scale. *Coral Reefs* 5:171–81.

Hubbard, D. K. 1986. Sediment as a control of reef development: St. Croix, USVI. *Coral Reefs* 5:117–25.

Hubbard, D. K. 1988. Controls of modern and fossil reef development: Common ground for biological and geological research. *Proceedings from the Eighth International Coral Reef Symposium* 1:243–52.

Hubbard, D. K. 1997. Reefs as dynamic systems. In *Life and death of coral reefs,* edited by C. Birkeland, 43–67. New York: Chapman and Hall.

Hubbard, D. K., J. D. Stump, and B. Carter. 1987. Sedimentation and reef development in Hawksnest, Fish and Reef Bays, St. John, U.S. Virgin Islands. *Biosphere Reserve Report* 21.

Hughes, T. P. 1989. Community structure and diversity of coral reefs: The role of history. *Ecology* 70(1):275–79.

Hughes, T. P. 1994. Catastrophes, phase shifts, and large-scale degradation of a Caribbean coral reef. *Science* 265:1547–51.

Hutchinson, G. E. 1953. The concept of pattern in ecology. *Proceedings of the National Academy of Science* 105:1–12.

Jackson, J. B. C. 1991. Adaptation and diversity of reef corals. *BioScience* 41(7):475–82.

Jackson, J. B. C. 1997. Reefs since Columbus. *Coral Reefs* 16, Supplement: S23–32.

Levin, S. A. 1992. The problem of pattern and scale in ecology. *Ecology* 73(6):1943–67.

Loya, Y. 1972. Community structure and species diversity of hermatypic corals at Eilat, Red Sea. *Marine Biology* 13:100–123.

MacArthur, R. H., and R. Levins. 1964. Competition, habitat selection, and character displacement in a patchy environment. *Proceedings of the National Academy of Science* 51:1207–10.

McManus, J. W., ed. 2001. *Priorities for Caribbean coral reef research: Results from an international workshop of the National Center for Caribbean Coral Reef Research (NCORE),* October 3–5, 2001, Miami, Fla. Retrieved from the National Center for Caribbean Coral Reef Research Web site: *www.ncoremiami.org/documents.html*

Murdoch, T. J. T., and R. B. Aronson. 1999. Scale-dependent spatial variability of coral assemblages along the Florida Reef Tract. *Coral Reefs* 18:341–51.

Rogers, C. S., V. Garrison, and R. Grober-Dunsmore. 1997. A fishy story about hurricanes and herbivory: Seven years of research on a reef in St. John, U.S. Virgin Islands. *Proceedings from the Eighth International Coral Reef Symposium* 1:555–60.

Rogers, C. S., L. N. McLain, and C. R. Tobias. 1991. Effects of Hurricane Hugo (1989) on a coral reef in St. John. *Marine Ecological Progress Series* 78:189–99.

Rogers, C. S., and R. Teytaud. 1988. Marine and terrestrial ecosystems of the Virgin Islands National Park and Biosphere Reserve. *Biosphere Reserve Report* 29.

Salm, R. V., J. Clark, and E. Siirila. 2000. Marine and Coastal Protected Areas: A guide for planners and managers. Washington, D.C.: The World Conservation Union.

Smith, S. V. 1978. Coral-reef area and the contributions of reefs to processes and resources in the world's oceans. *Nature* 273:225–26.

Sousa, W. 1985. Disturbance and patch dynamics on rocky intertidal shores. In *The Ecology of Natural Disturbances and Patch Dynamics,* edited by S. T. A. Pickett and P. S. White, 125–51. Orlando, Fla.: Academic Press.

Treml, E. 2000. Fringing reef framework development and maintenance of coral assemblages along the south shore of St. John, U.S. Virgin Islands. Master's thesis, University of Charleston, S.C.

Turner, M. G. 1989. Landscape ecology: The effect of pattern on process. *Annual Review of Ecology and Systematics* 20:171–97.

Turner, S. J., R. V. O'Neill, and W. Conley. 1991. Pattern and scale: statistics for landscape ecology. In *Quantitative methods in landscape ecology,* edited by M. G. Turner and R. H. Gardner, 17–49. New York: Springer-Verlag.

Urban, D. L., R. V. O'Neill, and H. H. Shugart. 1987. Landscape ecology. *BioScience* 37:119–27.

USCRTF. 2000. National action plan to conserve coral reefs. United States Coral Reef Task Force. Washington, D.C.

Watt, A. 1947. Pattern and process in the plant community. *Journal of Ecology* 35:1–22.

Wiens, J. A. 1989. Spatial scaling in ecology. *Functional Ecology* 3:385–97.

Wilkinson, C. 2000. Executive summary. In *Status of coral reefs of the world: 2000,* edited by C. Wilkinson, 7–20. Queensland, Australia: Australian Institute of Marine Science, Cape Ferguson.

Wischmeier, W. H. 1976. Use and misuse of the universal soil loss equation. *Journal of Soil and Water Conservation* 31:5–9.

Witman, J. D. 1992. Physical disturbance and community structure of exposed and protected reefs: A case study from St. John, U.S. Virgin Islands. *American Zoologist* 32:641–54.

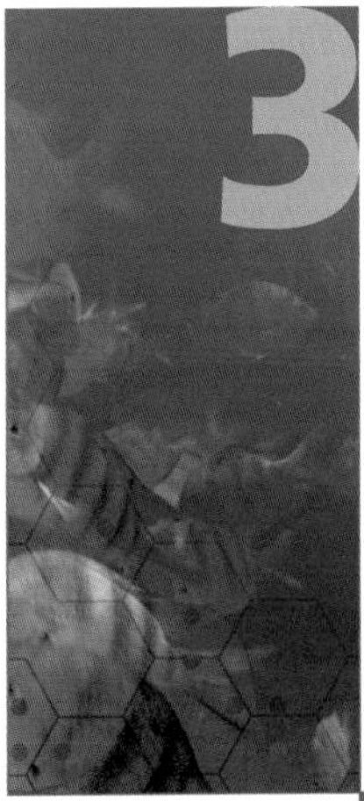

3 Green Sea Turtles Migration in the Gulf of Mexico

Satellite Telemetry and GIS

Michelle Rene Kinzel
Oceanic Resource Foundation
San Francisco, California

During August and September 2000, on an isolated, subtropical beach in Veracruz, Mexico, a small team of researchers, sea turtle biologists, and local volunteers put into action a satellite telemetry project, SAT TAG 2000, that had been months in the planning. In an effort to promote international conservation and support research efforts in Mexico and the United States, the Oceanic Resource Foundation secured funding and support for this satellite telemetry study. SAT TAG 2000 was established and conducted in response to the urgent need to develop international measures of conservation for all species of sea turtles, as well as their habitats.

Turtles have the innate navigational ability to return to their beach of origin periodically throughout their lives.

PHOTO COURTESY OF MICHELLE KINZEL

The objectives of this study were to determine the migratory corridors and habitat usage patterns of green sea turtles, *Chelonia mydas,* that nest on the beaches of Lechuguillas, Veracruz. The long-term goals of this study included providing the scientific findings to conservationists and policy makers in an effort to increase protection of this endangered species and its crucial habitats. GIS plays an important part in this effort.

PHOTO COURTESY OF MICHELLE KINZEL

Newly hatched turtles emerge from their nests under the sand.

Sea turtles are mysterious and alluring sea creatures that have inhabited the earth for millions of years and have been termed living dinosaurs by sea turtle biologists. Sea turtles have also been referred to as ambassadors of the sea because of the integral roles they play in the ocean ecosystems of multiple nations during their life cycles. Due to the migratory nature of sea turtles, their habitats are varied and wide-reaching as they often traverse expansive ocean basins and occupy the territories of many sovereign nations. Most species will nest on the beaches of one nation and then travel across oceans or international borders to feed in another. These ancient reptilian species are what scientists refer to as "keystone species." This term refers to the sea turtles' functional role within their ecosystem. Their feeding habits and actions affect organisms lower in the food chain. The sea turtle provides a focal point upon which to initiate ecosystem conservation and recovery programs.

There are six species of sea turtles that inhabit the Pacific and Atlantic oceans off the United States: the green turtle *(Chelonia mydas)*, hawksbill *(Eretmochelys imbricata)*, Kemp's ridley *(Lepodochelys kempii)*, leatherback *(Dermochelys coriacea)*, loggerhead *(Caretta caretta)*, and olive ridley *(Lepodochelys olivacea)*. At present, all six species are protected under the Endangered Species Act and listed as either threatened or endangered by the U.S. Fish and Wildlife Service. All but the Kemp's ridley turtle migrate to Mexican waters to reproduce and deposit their eggs along sandy beaches (Bravo 1995).

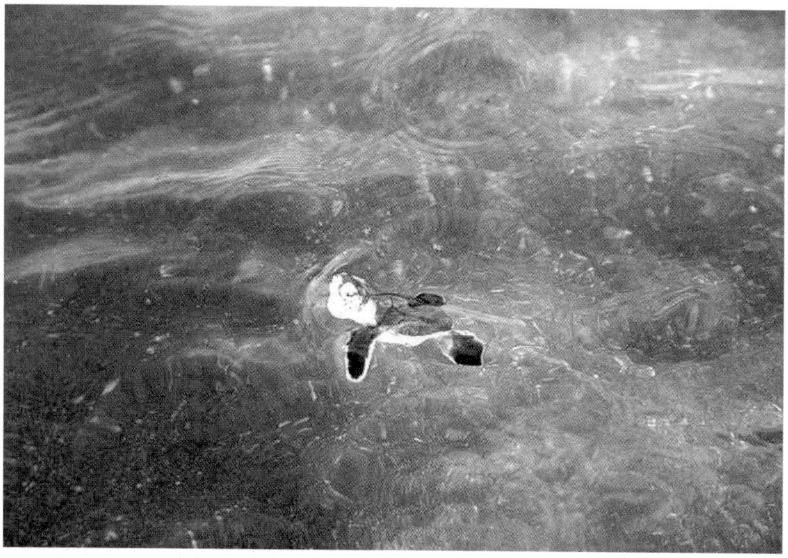

PHOTO COURTESY OF MICHELLE KINZEL

Most at home in the water, turtles can hold their breath for more than 15 minutes and migrate across wide regions.

Sea turtles face an onslaught of threats during every stage of their life cycle. Female sea turtles spend nearly all of their time in the ocean, returning to the terrestrial environment only briefly as they lay their eggs to propagate the next generation (Musick and Limpus 1997). This journey onto land is a brief event but one that makes the turtles susceptible to human poachers and predators, including the jaguar. The sea turtles' reliance on a terrestrial nesting site and return to their natal beaches makes them vulnerable to being caught and killed for their meat, shells, and eggs.

The females do not incubate the nests, choosing instead to roll the dice in a biological reproductive numbers game that will have the females laying up to seven hundred eggs per breeding season. Thus, the eggs are highly vulnerable and sought by a myriad of predators including feral dogs, nocturnal mammals, beach crabs, and man (Musick and Limpus 1997).

Both male and female adult turtles are easily caught in the open ocean when they surface to breathe. Also, a significant cause of mortality is from incidental by-catch of several coastal fisheries. Add to these factors commercial development along nesting beaches, destruction of foraging habitats, ingestion of plastic debris, entanglement in floating debris, boat propellers, toxic pollution, and lighting disorientation of hatchlings, and it is easy to see that these slow-maturing animals face a perilous and treacherous 25- to 30-year struggle to reach sexual maturity (Musick and Limpus 1997).

Nesting beach conservation efforts began in earnest in Mexico in 1964, and organized protection of sea turtle populations and nesting beaches has been occurring ever since. Several environmental and research organizations monitor and protect reproducing female turtles and their nests (Bravo 1995). The

Mexican government has been instrumental in the establishment and funding of several such organizations. Despite the support and conservation efforts of the United States and Mexico, however, sea turtle populations are either declining in numbers or only marginally recovering (Bravo 2000).

The mature turtle with satellite transmitter on her back heads to sea.

PHOTO COURTESY OF MICHELLE KINZEL

Recent increased protection has been granted to sea turtle species along their migratory routes, because of severe population losses caused by incidental by-catch in fishing nets. The U.S. government has mandated the closure of shrimp fisheries during the crucial migratory season, March and April, along the coastal waters of Texas. In addition, because U.S. fisheries have been identified as impacting sea turtle species, the U.S. Commerce Department mandates that commercial fisheries use Turtle Excluder Devices when harvesting the seas (Epperly 1995). The data from projects such as SAT TAG 2000 helps conservation managers make sound recommendations for the management of fisheries in areas inhabited or utilized by endangered sea turtle populations.

To effectively protect sea turtle populations and implement conservation efforts, information on feeding ground locations, nesting beach occupancy patterns, and migratory routes is needed. By employing GIS to spatially depict sea turtle occupancy and patterns of usage, biologists can better assess potential threats to these animals and propose effective conservation measures to eliminate or reduce these threats.

Historically, this information has been sparse and collected by monitoring the females and eggs during their brief residences on nesting beaches. Previous biological studies have relied on monitoring adult female turtles by attaching metal or plastic flipper tags, injecting Passive Integrated Transponder (PIT) tags into flipper muscle, or creating "living tags" (Balasz 1999). The flipper tags are cost effective, while the PIT tags involve the injection of a microcomputer chip and require expensive equipment both to insert and later read the bar code. The microprocessor chip has an individual bar code, and researchers must use electronic scanners to read the tags. This method, as well as the creation of living tags, is costly and relatively time consuming.

These methods essentially mark the turtle and can provide information regarding age classes, transoceanic or long-range movements, nesting frequency, and repetitive visitation to the nesting beaches. However, these methods of identifying and studying the turtles do not provide the detailed information on migratory routes, feeding site occupancy, or habitat usage that is crucial to conservation efforts.

PHOTO COURTESY OF MICHELLE KINZEL

Newly hatched turtles emerge from the sand to enter the ocean for the first time.

The modern era of microelectronics has provided unique opportunities to study sea turtle movements and behaviors at sea. Sea turtle biologists began using satellite telemetry technology in the 1980s. Early tracking studies used the Nimbus climatological system and provided data on the movements of loggerhead turtles in the Georgia Bight (Stoneburner 1982, Timko and Kolz 1982). The first confirmation of a trans-Pacific movement by a loggerhead turtle was made from the recovery of a flipper tag (Resendiz et al. 1998). This information, while enlightening, was limited in detail to the

Estimated home range of Roberta shown in overlay of benthic habitat. The habitat is largely covered by patchy or continuous sea grass beds.

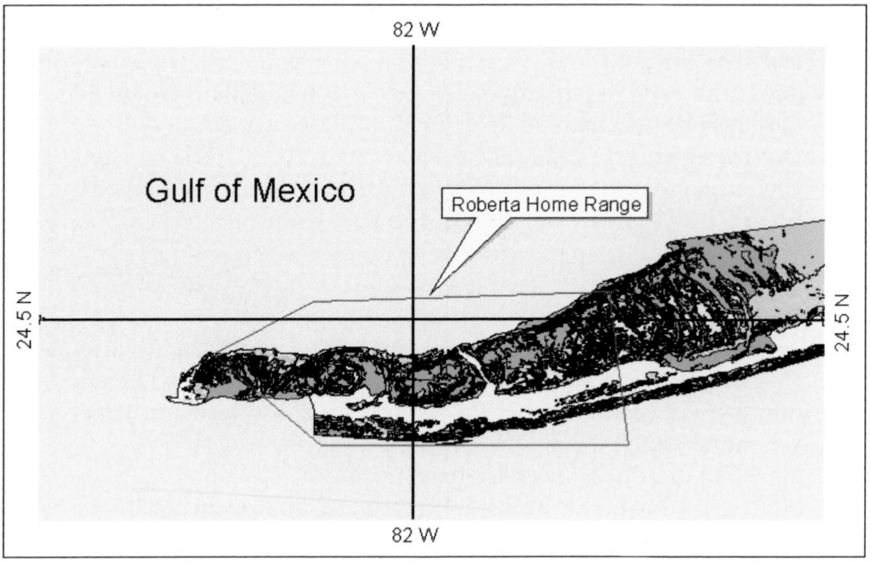

Area of Selected Polygon: 2745.63 sq kilometers

Bare Substrate: 39.9325 sq kilometers / 01.5%
Continuous Seagrass: 563.3631 sq kilometers / 20.5%
Hardbottom: 152.0012 sq kilometers / 05.5%
Inland Water: .5858 sq kilometers / 00.0%
Land: 86.1531 sq kilometers / 03.1%
Patch Reefs: 20.2912 sq kilometers / 00.7%
Patchy Sea Grass: 406.1419 sq kilometers / 14.8%
Platform Margin Reefs: 70.0159 sq kilometers / 02.6%
Unclassified: 1012.8233 sq kilometers / 36.9%
Unknown Bottom: 394.326 sq kilometers / 14.4%

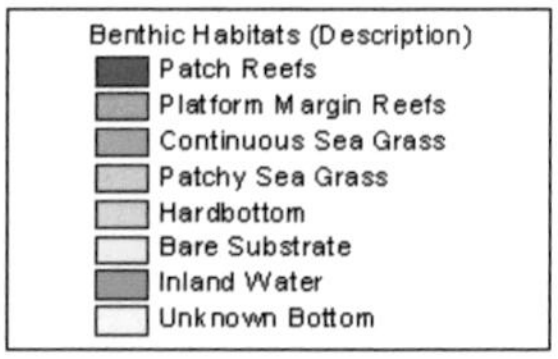

starting and ending points of the migratory route. By using satellite tags, one researcher was able to elucidate an important part of the migratory puzzle for the scientific community. In 1996, Wallace J. Nichols was able to confirm the migratory route from Baja California to Japan via a turtle tracked with satellite telemetry (Nichols et al. 1998). Nichols was able to determine exact routes of migration and swim speeds with the use of this new technology.

Two sexually mature female sea turtles were chosen for this study. The female turtles were found during a night patrol of their nesting beach and were detained following the deposition of their egg clutches, which were safely transported to a protected and monitored corral. The turtles were placed in wooden boxes to minimize their movements as researchers attached the small transmitters to the carapace (or hard shell) using a silicone polymer. The transmitting devices were secured with fiberglass strips and painted with an olive-colored marine paint to help camouflage the units with the carapace. The females, Zyanya and Roberta, having been tagged with steel Monel flipper tags, embarked on a journey that would contribute to our sparse but quickly growing knowledge of green sea turtles and their movements between natal nesting beaches and feeding grounds.

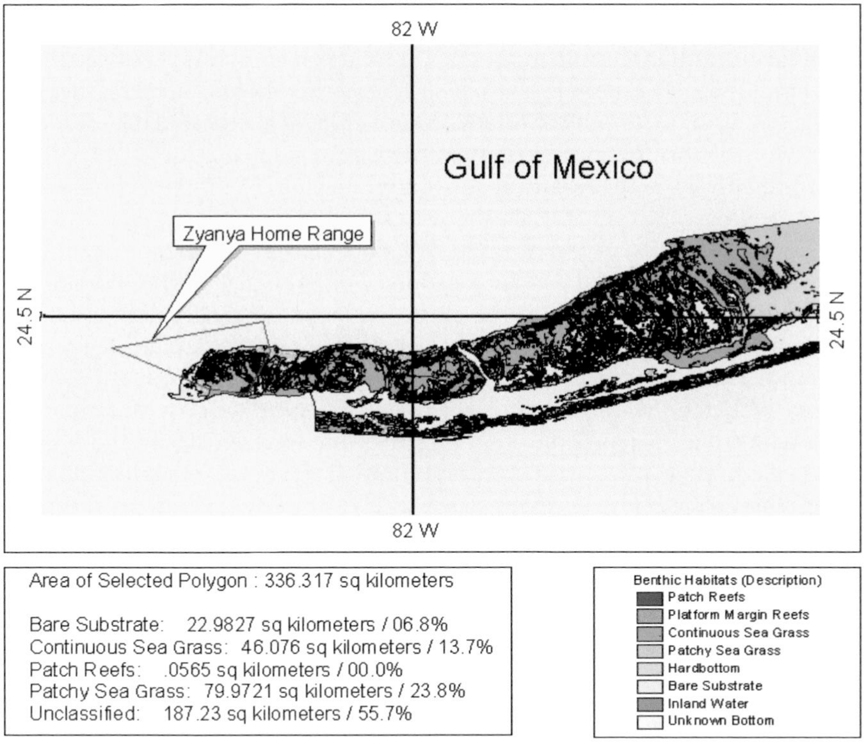

Estimated home range of Zyanya shown in overlay of benthic habitat. The habitat is notably smaller than the range of the other turtle.

Even before Zyanya and Roberta were released, the tracking process began. The transmitting units measure 13 cm × 4 cm × 2 cm, approximately the size of a television remote-control device, and are small relative to the size of the turtles' carapaces, which measure approximately 1 meter long. Each transmitter contains a microprocessor, which is preprogrammed with a duty cycle chosen by the researchers. In order to maximize duration of transmissions, the transmitters are programmed to send their signals using low-wattage output, ranging from 0.5 to 1 watt, for a predetermined duty cycle, which is followed by an off-duty cycle. For this study, a duty cycle of six hours on, followed by 18 hours off, was chosen.

The transmitters send their signal, which is detected by one of the four polar-orbiting ARGOS satellites that have been designated for tracking wildlife. Satellites receive data during their overpass in the region of the transmissions. Each satellite circles the earth every 101 minutes, and the satellites are in position to receive data from any one location for approximately 10 minutes (ARGOS 1990).

Because the quality of signal transmission is dependent on the angle of the satellite over the horizon and the number of overpasses in one location, the best readings are collected near the poles. Data collected from the equator has a higher frequency of lower-location class readings.

To determine the location of the transmitter, a satellite must receive multiple readings from the transmitting unit. An average reading takes between three and five minutes. This can prove to be problematic with an aquatic species such as the green sea turtle, which usually surface only to breathe and do not remain at the surface for extended periods of time. As a result, readings are received intermittently, with the results of this study showing data input approximately every three days (Kinzel 2001).

The signals are categorized into location classes based on the number of readings from that particular location. The accuracy of the latitude and longitude position readings can be estimated based on the location class. The location classes, from poorest to best, include Z, B, A, 0, 1, 2, and 3. Location classes of Z indicate that the transmitter is sending a signal but the location cannot be determined. Therefore, readings from location class Z are rejected for tracking and mapping purposes. Signals from location classes B and A are somewhat reliable, but they are not as accurate as the classes zero through three. Researchers prefer to use location classes of zero or greater, which are estimated to be within 350 meters of accuracy, for plotting maps and analyzing rates of travel. For the two turtles tagged in this study, the majority of location classes were rejected because only 21.21 percent of the readings for Zyanya, and only 18.92 percent of the readings for Roberta, were of location class zero or greater. Zyanya covered a route of 985 kilometers, at an average rate of .631 km/hr. Roberta traveled a bit slower, at .362 km/hr for a total of 1,430 kilometers.

PHOTO COURTESY OF MICHELLE KINZEL

The satellite transmitter remains intact on the turtle's carapace (hard shell), enabling the satellite to track her migration.

With the recent accessibility and implementation of satellite technology, more detailed information is now available to biologists, conservation managers, and policy makers. Satellite tracking allows for data retrieval via a transmitted signal and does not rely on the recovery of the tagging instrument. By combining the application of a transmitting device to the body of an animal with a means of signal retrieval, such as an ARGOS satellite that orbits the earth, researchers can track the specific movements of the animal over long distances.

This field project is a cooperative effort between Oceanic Resource Foundation of the United States and the Mexican environmental organization Centro Regional de Investigación Pescevera-Instituto Nacional de la Pesca en Veracruz (CRIP–INP–VER). Researchers have successfully attached ST-18 Telonics transmitters to two female turtles at the end of their nesting and egg-laying season. ARGOS satellite data was transmitted between September 2000 and April 2001. GIS analysis has been used to plot the latitude and longitude readings into map tracks of the migratory routes, assess home ranges, and analyze habitat utilization.

The two female sea turtles followed different routes in their movements across the Gulf of Mexico. Roberta remained close to the nesting beach for 10 days before traveling south and crossing the Gulf of Mexico near the northern end of Cuba. Zyanya sent signals from locations on or near the nesting beach for 21 days before heading north and remaining in relatively shallow water near the coast before heading across the Gulf of Mexico near southern Texas. Both tagged turtles have taken up residence in feeding grounds located within the Marquesas Keys, near the Florida Keys in the Gulf of Mexico.

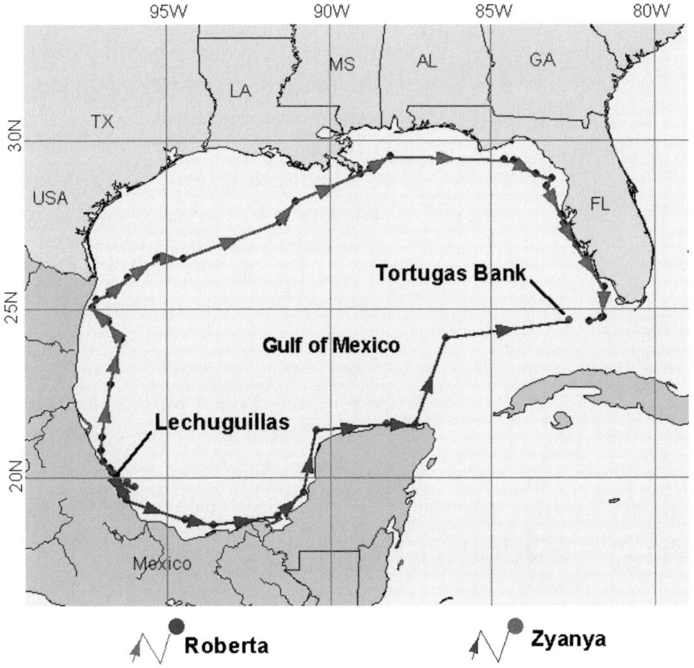

Detailed movements of green sea turtles Roberta and Zyanya, from the nesting beach in Lechuguillas, Mexico, to the foraging grounds off the Florida Keys. Data was collected via satellite telemetry using the ARGOS satellite system.

Latitude and longitude data from the ARGOS satellites was analyzed using GIS analysis and maps of the Florida Keys. The turtles' transoceanic movements and home ranges were analyzed using ArcView 3.2 and an ArcView extension entitled Animal Movement 2.04, created by Philip Hooge of the U.S. Geological Survey (USGS), Alaska Biological Center. The home range for each turtle was estimated using Animal Movement's minimum convex polygon function. The home ranges were plotted over the benthic substrates with the use of "Benthic Habitats of the Florida Keys," a CD from the National Oceanic and Atmospheric Association (NOAA), and ArcView Data Publisher.

The home range for Roberta was estimated to be 2,745.63 square kilometers. Zyanya's home range was reported as being 336.317 kilometers. The smaller estimated home range for Zyanya is most likely a factor of a small data set and poor ARGOS satellite latitude/longitude class readings. The final transmission from Zyanya's transmitter was logged on March 22, 2001, while Roberta's readings continued through April 18, 2001. The home range reported for Roberta is therefore assumed to be more accurate in terms of actual habitat usage, and not useful for estimating total occupancy in terms of square kilometers. The analysis of both home ranges for occupancy in specific substrates indicated the turtles are maintaining residence in areas of high sea-grass concentration. The combined categories of continuous and patchy sea grass totaled 35.3 percent for Roberta and 37.5 percent for Zyanya. These findings are consistent with previous research that found sea turtles to be highly selective feeders. Researchers have discovered microbial and ecological evidence that indicates these herbivorous reptilians may be grazing in the same location, recropping the same plots of sea grass.

The use of ArcView and extensions, such as Animal Movement, can have profound and far-reaching implications for conservation. Ecological studies and analysis of animal behavior have expanded their focus to include detailed examination of habitats. Knowing the threats to a habitat and the impact on the animals is key to any conservation effort. Sea turtles face many threats while occupying their foraging grounds, including marine debris ingestion, commercial fishery interactions, diminishing food resources, damage to flora by anchors and propellers, deposition of silt from land runoff, and oil spill catastrophes. Integrating the capabilities of GIS analysis will help elucidate these factors and point the way toward effective conservation measures.

Acknowledgments

SAT TAG 2000 was accomplished with the efforts of many people and organizations. Greg Carter of the Oceanic Research Foundation initiated the international project in collaboration with the Mexican organization Centro Regional de Investigación Pescevera–Instituto Nacional de La Pesca en Veracruz (CRIP–INP–VER), and marine biologists from Mexico and the United States. Graciela Tiburcio Pintos served as a biologist on the project and coordinated operations and fieldwork in Mexico. Rafael Bravo Gamboa, an engineer with CRIP, served as camp manager and coordinated all logistical operations at the tagging site. The author served as a biologist during fieldwork. Greg Carter monitored the ARGOS satellite data and prepared the maps of the turtles' oceanic movements and home range estimates, using ArcView and Animal Movement software programs. Additional information about SAT TAG 2000 and these magnificent creatures can be found at *www.orf.org/turtles.html.*

PHOTO COURTESY OF MICHELLE KINZEL

A satellite transmitter fastened to the turtle's carapace with epoxy resin does not injure the turtle, but enables the satellite to track the turtle's migration.

References

ARGOS. 1990. *User's manual.* Landover, Md.: Service Argos, Inc.

Balazs, G. H. 1999. Factors to consider in the tagging of sea turtles. In *Research and management techniques for the conservation of sea turtles,* IUCN/SSC Marine Turtle Specialist Group Publication No. 4, edited by Karen Eckert, et. al., 101–9.

Bravo, G. P. R., I. T. Hernandez, and A. P. Pech. 2000. Protección y conservación de las tortugas marinas en playas de Lechugillas, Mpio. de Alatorre, Ver. Durante la temporada de anidación 1999. Informe Técnico Interno Anual. SEMARNAP–INP–CRIP.

Bravo, G. P. R. 1995. Investigación y protección sobre tortugas marinas que anidan desde playas de Navarro hasta Santa Ana, Mpio. De Alto Lucero, Veracruz, Mexico. Temporada 1994. Informe Técnico Interno Anual. SEMARNAP–INP–CRIP.

Epperly, S. P., J. Braun, and A. Veishlow. 1995. Sea turtles in North Carolina waters. *Conservation Biology* 9 (2):384–94.

Kinzel, M. R. 2001. Satellite tracking of green sea turtles in the Gulf of Mexico. *ARGOS Newsletter* No. 58.

Musick, J. A., and C. J. Limpus. 1997. Habitat utilization and migration in juvenile sea turtles. In *The biology of sea turtles,* edited by P. L. Musick and J. A. Lutz, 107–36. Boca Raton, Fla.: CRC Press.

Nichols, W. J., et. al. 1998. Using molecular genetics and biotelemetry to study sea turtles migration: A tale of two turtles. *18th Annual Symposium on Sea Turtle Biology and Conservation.* Mazatlan, Mexico.

Resendiz, A., et. al. 1998. First confirmed east–west transpacific movement of loggerhead sea turtles *(Caretta caretta),* released in Baja California, Mexico. *Pacific Science* 52(2):151–53.

Stoneburner, D. L. 1982. Satellite telemetry of loggerhead sea turtle movement in the Georgia Bight. *Copeia* 1982(2):400–408.

Timko, R. E., and A. L. Kolz. 1982. Satellite sea turtle tracking. *Marine Fisheries Bulletin* 44:19–24.

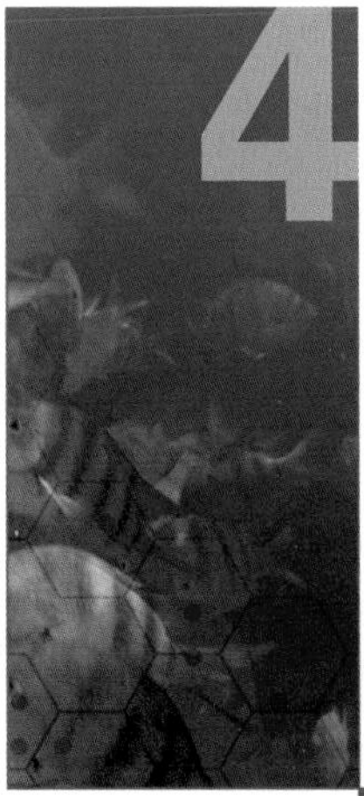

Marine GIS for Management of Scottish Marine Special Areas of Conservation

Richard Bates
School of Geography and Geosciences, University of St. Andrews
St. Andrews, Scotland

Ben James
Scottish Natural Heritage
Edinburgh, Scotland

In May 1992, the European Community published the Habitats Directive aiming to conserve internationally important species and habitats across their European range. The first part of the directive required member states to identify, designate, and protect the most important sites to ensure that the extent and range of specific habitats and populations of constituent species are maintained over time within these Special Areas of Conservation (SAC). The Habitats Directive also enabled member states to establish management plans for the long-term maintenance of the SACs. The sites included both terrestrial and marine habitats and species. The human activities taking place within the sites were also considered where management measures were enacted, and focus on the conservation interest was maintained. Consideration of the sustainable development of human activities within marine SACs marked this directive as different from many historical conservation actions and demonstrated that the conservation of species diversity is not seen as incompatible with sustaining human activities within protected areas. The use of GIS became a key component in the mapping and management of marine SACs along with the use of remote survey and evaluation techniques similar to those provided by geophysical sonar that is now widespread in the marine community.

In 1996, the U.K. marine conservation organizations established the UK Marine SACs Project to help implement the Habitats Directive in the United Kingdom. The key tasks of this project included the development of methods to enable an assessment of the condition of habitats within 12 marine SACs around the coast of Britain and the management of these sites. A key message to emerge from the Marine SACs Project was the need to have a process for managing and integrating the large and diverse databases of information that exist for each site. It was also essential that this information be disseminated to a wide audience of stakeholders (English Nature et al. 2001). The Habitats Directive also requires member states to undertake surveillance on the condition of the sites with the subsequent reporting of this work every six years (articles 11 and 17). The use of GIS is core to meeting these obligations, in particular those of initial marine SAC surveillance, mapping the distribution of individual habitat components, and establishing baselines for determining future change. In Scotland, 34 marine sites were proposed for SAC designation and the acquisition of the necessary data for management purposes for each of these is ongoing under the overall direction of Scottish National Heritage.

Scottish marine SACs

The coastline of Scotland, at nearly 16,500 kilometers in length, contains a vast range of habitats from flooded fjords to enclosed bays and large firths or estuaries. The west coast is highly exposed to the North Atlantic, but has a climate that is modified by the North Atlantic Drift. The character of the shoreline is dependent on a number of physical factors including wave and wind exposure, salinity, geology, tidal range, and the strength of currents. More than 85 percent of Scotland's population lives within 10 kilometers of the coast and a significant proportion of Scotland's revenue is derived from activities conducted in the coastal zone. As a consequence, it is vital to the long-term sustainable development of the coastal zone that the natural habitats are preserved and managed in conjunction with economic development.

Each marine SAC around Scotland's coast is unique with different biological, physical, and human elements. In order to manage each of the sites, it is necessary to not only evaluate the key elements such as the species, setting, and human activity, but also to understand why these components are important. It is only through this process that one can then evaluate the level of change on a site that might have a detrimental impact upon the important habitats of species and cause site degradation. Questions about the future use of these sites must be addressed by local user groups, conservation organizations, and government to ensure that their management is to the long-term benefit of society. For each of the SACs, a thorough evaluation of the scale of natural "inherent" variability within the system is undertaken so that impacts of an anthropogenic nature can be determined. Changes within marine SACs could include physical modifications such as an increase in siltation (e.g., from increased dredging activity), increases in abrasion from more frequent storm activity, and chemical changes arising from synthetic and nonsynthetic compounds or radionuclides.

Survey program

A variety of remote and direct sampling methods is necessary for the collation and evaluation of historic information and the acquisition of new data for each marine SAC (Foster-Smith et al. 2000). Broad-scale remote sampling methods include satellite observations, aerial photography, and acoustic surveys. Satellite and aerial photography techniques have limited water penetration capabilities within the marine environments on the east coast of Scotland, but are more applicable in the clearer waters on the west coast where light can penetrate 5 to 15 meters. These methods provide rapid, large-area coverage for some of the diverse shallow inlets, bays, and marsh habitats that would be prohibitively expensive to survey using other techniques. For rapid broad-scale remote mapping of areas not accessible to aerial remote-sensing techniques, acoustic methods based on echo sounders, sidescan sonar, and multibeam sonar are used.

Both the aerial and acoustic remote survey techniques require "ground truth" validation using data collected by a range of standard sampling methodologies.

Over the last two years, procedures have been developed by the University of St. Andrews and Heriot Watt University that enable a rapid turnaround in the processing of broad-scale acoustic survey information. All geophysical survey data was acquired and processed on-site during the survey to provide digital bathymetric models with draped sidescan and classed sonar reflections from the seafloor. During data collection, a separate computer running ArcView is present on board the survey vessel so that data can be transferred via a network in near real time. This computer is also in communication with the navigation computer providing navigation information from a differential global positioning system (GPS) *(figure 1).* Preliminary biological evaluation and sediment classification is accomplished in the field based on information derived from a range of sampling techniques including diver-based observations with video and still cameras, remote video drops, ROV (remote operated vehicle) video, and sampling from grabs and cores. The seafloor habitat type was initially classed by combining this ground validation data with a knowledge of species biology, bathymetric position (aspect and slope), sidescan textural signature, and acoustic ground discrimination values. The habitats are grouped together in areas or polygons that exhibit the same ranges of conditions. At present, this process is a manual comparison of conditions with limited discrimination scripts written in the Avenue™ programming language; however, current research is focusing on the development of a set of logic-based automated habitat assignment routines for use within the GIS.

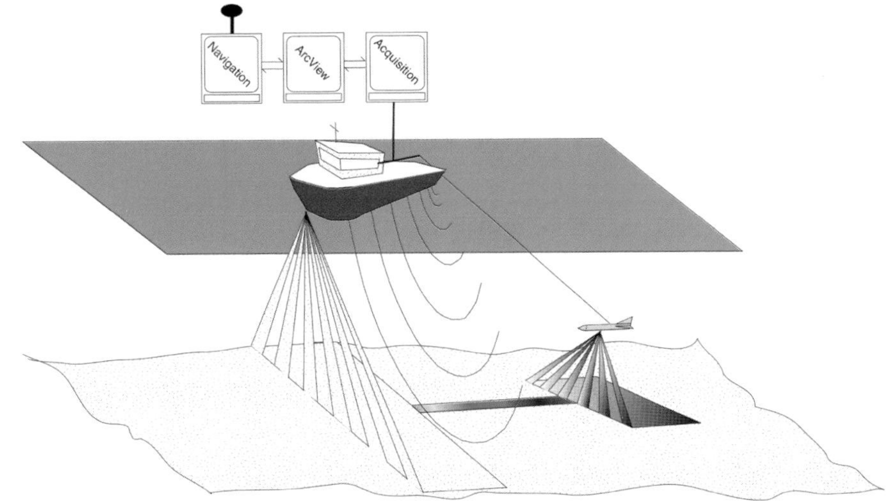

Figure 1: Schematic of acquisition and on-board processing system.

Historical data

An evaluation of historical data for input to the GIS is vital if the database is to be used as a management tool. In many SAC areas, the first novel survey data collected might be regarded as the baseline data; however, there is often extremely important information in the recent geological record that should also be considered. For example, an evaluation of changes in the Scottish landscape is not complete without consideration of the changes in agricultural practice during the highland clearances of the 1700s to 1800s. During this time, when large-scale sheep-farming practices were introduced and local deforestation began, there was a significant change in physical and

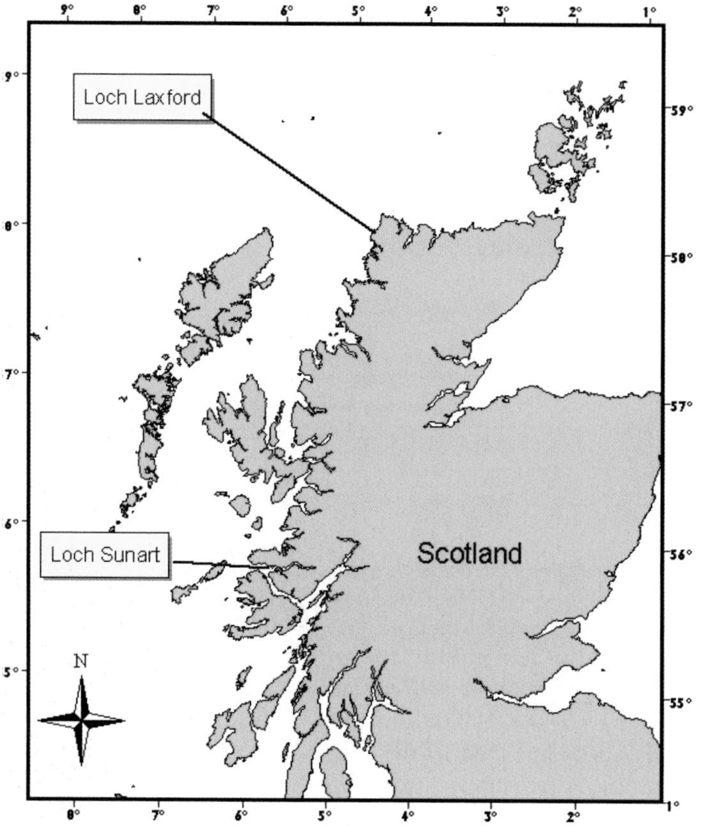

Figure 2: Location map of Loch Laxford and Loch Sunart marine Special Areas of Conservation.

chemical inputs from streams to coastal areas. These changes are recorded in the sediment accumulations within lochs as is the story of climatic change during the last glaciations. Coring programs for sampling and dating subbottom material from the sea lochs are an important part of the baseline data gathering from these areas of conservation.

New survey data

A detailed bathymetric chart is a key element to understanding and delineating the range of habitats within marine SACs. Historically, these have been provided by access to U.K. Admiralty Charts or by taking echosounding measurements in the field. In these formats, the spatial resolution is not sufficient to record the small-scale changes that can influence the biological communities encountered. Therefore, it is necessary to obtain new bathymetric information for the construction of digital models using sidescan systems such as the Submetrix System 2000 and multibeam sonar such as the Reson Seabat 8125. Information from these detailed charts includes not only the depth but also the slope and slope angle or exposure angle of the seafloor. This is of particular importance when determining the spatial distribution of marine benthic communities.

The benefit of this high resolution for the undertaking of a geological appraisal is demonstrated in Loch Sunart, a 28-kilometer-long and 130-meter-deep sea loch on the west coast of Scotland *(figure 2).* The bathymetric profile of the loch is complex as it is subdivided into three major basins or subloch units separated by narrow and shallow sills. Each of the subbasins contains remnant features from the last major glacial events in northern Europe during the Devensian period (before approximately ten thousand years ago) and sedimentological features resulting from deposition during glacial retreat. At the eastern end of the inner loch, these include a number of approximately north-to-south-trending, sharp-crested ridges identified on the bathymetric chart *(figure 3a).* The amplitude swaths (sidescan-like images) over these features suggest that they are sediment ridges rather than exposed bedrock at the surface *(figure 3b).* An analysis of the slope magnitude *(figure 3c)* and aspect of the ridges *(figure 3d)* using GIS show that the west-facing slopes were steeper than east-facing slopes.

A comparison of the shape of these features with other glacial moraines reveals that they could represent De Geer-type glacial moraines similar to those reported on Baffin Island by Boulton (1986) for debris deposited by grounded ice in front of glaciers. These features are not evident in the deeper parts of the loch below 40 meters. This suggests a rapid ice retreat across the over-deepened section leading to grounding and stalling of the retreating ice sheet on the shallower seafloor to the east. In order to determine the exact nature of these features, a follow-up research program is proposed using sub-bottom profiling systems to determine the stratigraphy of the cores.

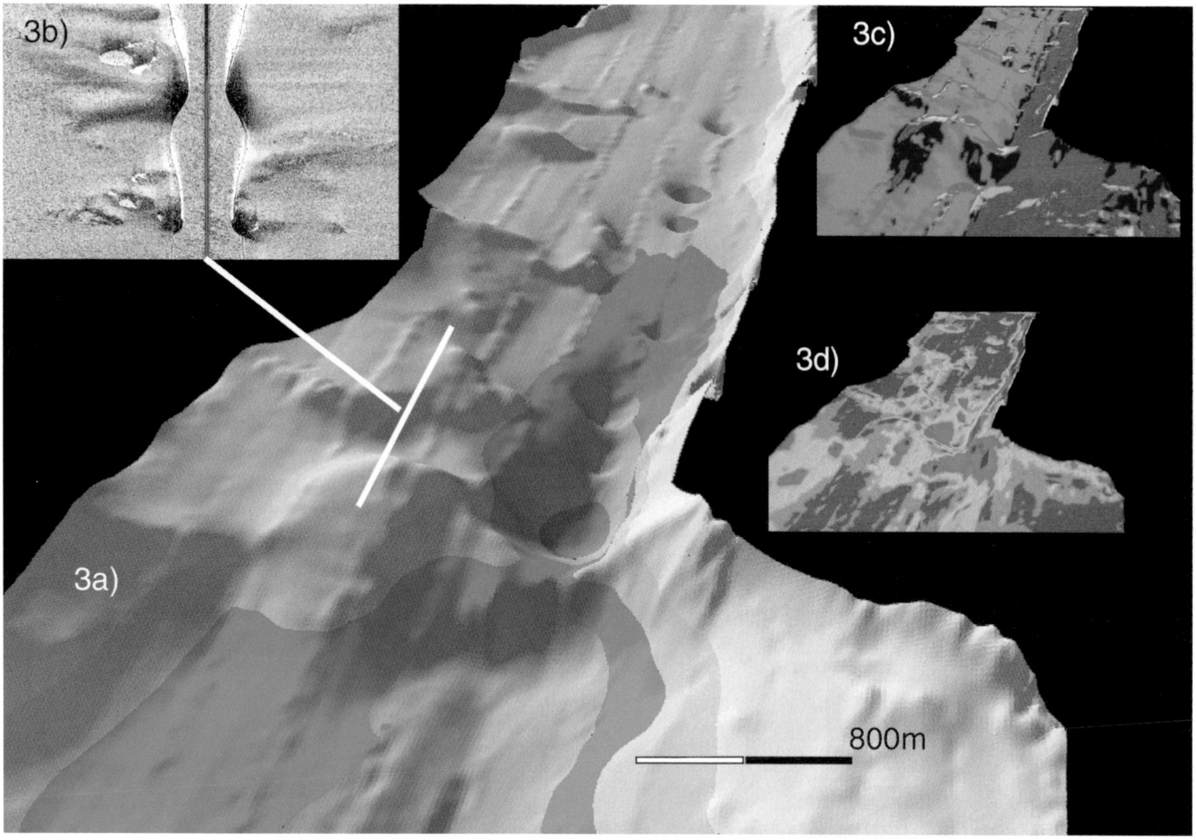

Figure 3: (a) Digital bathymetric model of inner Loch Sunart; (b) amplitude (sidescan) image over north-to-south-trending ridges; (c) slope aspect across ridges; (d) slope magnitude across ridges.

Surface information

Detailed scientific information on the nature of the seafloor is available through a number of acoustic remote-sensing methods including sidescan sonar, bathymetric sidescan, multibeam sonar, and acoustic ground discrimination sonar. The resulting images are then draped onto the bathymetry (Blondel and Murton 1997). While this provides a good representation of bottom conditions, the method often suffers from difficulty in georeferencing the images. With the new generation of multibeam sidescan, bathymetric sidescan, and multibeam sonar, georeferencing is achieved during the acquisition of the data. The resulting amplitude maps have very high spatial resolution (Bates and Byham 2001). A view of an amplitude map draped on a bathymetric chart for Loch Laxford *(cf. figure 2)* on the west coast of Scotland is shown in figure 4a. The spatially referenced view clearly shows the rocky outcrops as the dark areas of high bathymetric relief and the uniform sediment in the bottom of the loch as lighter shades. This interpretation is further corroborated with information on the strength of acoustic echo-signal reflection from the seafloor recorded using Acoustic Ground Discrimination Sonar (AGDS). AGDS reflection strength in terms of the roughness and hardness of the seafloor is compared to biological and sedimentological type through ground-truth observations (Chivers et al. 1990; Sotheran et al. 1997). The method has the advantage of being easily implemented on a wide range of survey vessels and records data along line tracks from beneath the survey vessel. These have been color-coded as an overlay to the sidescan over the rock areas shown in detail in figure 4b.

Figure 4: (a) View looking east into the outer part of Loch Laxford; (b) expanded view of loch floor overlaid with acoustic ground discrimination line track data.

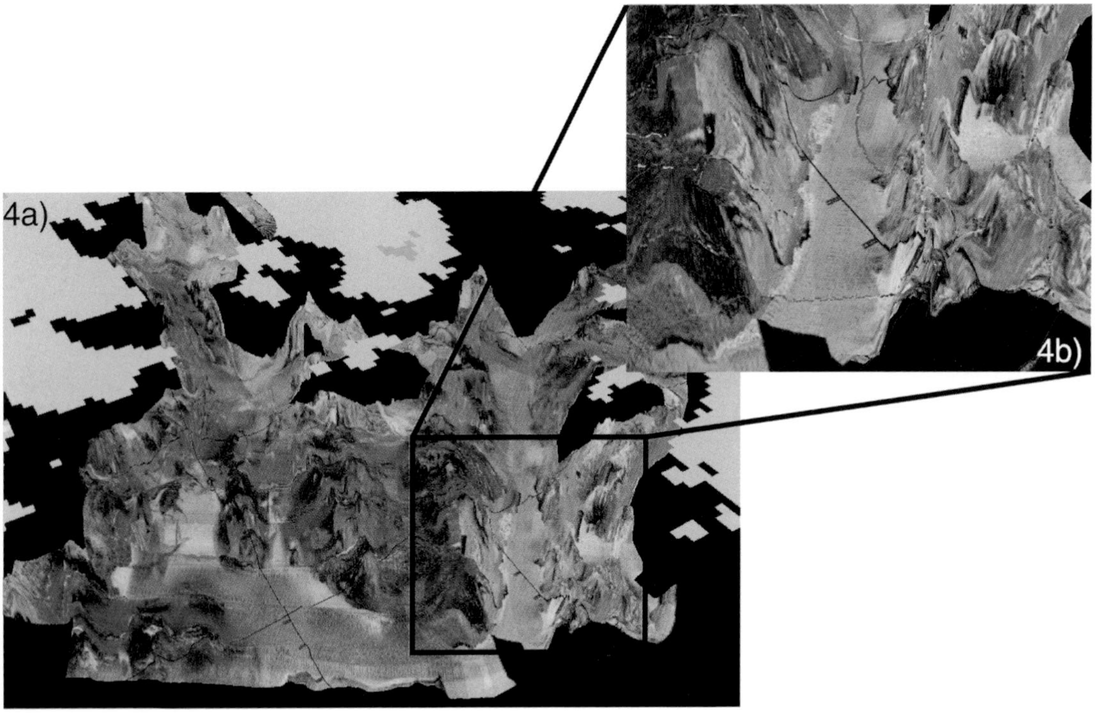

Future developments

While the information described above provides a detailed broad-scale description of benthic conditions within a marine SAC, to date information is not regularly recorded for properties in the water column at the sites. Data about the volume of water and the water quality within a marine SAC also provides critical information about the health of the sites. Evaluating the water column is more problematic than evaluating signatures on the seafloor, as changes tend to occur within a more rapid time frame and thus require the GIS to include time-based data. Key information from the water column includes temperature, salinity, conductivity, currents, and chemical signatures. This information is used for creating many of the circulation models and, in conjunction with the data from sub-bottom sediments, for re-creating circulation models for past climatic conditions. At present, this type of analysis is at the research stage in the School of Geography and Geosciences for the Scottish marine SACs program and many challenges are foreseen with respect to addressing true volumetric measurements of the water column. In assessing transient populations within the marine SACs (pelagic fish and marine mammals), similar difficulties are expected. Direct sampling and acoustic measurements are marginally successful in evaluating these aspects of an SAC. However, it is anticipated that the latest technologies such as the multibeam sonar may provide a way forward and represent the next generation of survey technologies.

References

Bates, C. R., and P. Byham. 2001. Swath-sounding techniques for near-shore surveying. *The Hydrographic Journal* 100:13–18.

Blondel, P., and B. J. Murton. 1997. *Handbook of Seafloor Sonar Imagery.* Chichester: John Wiley-Praxis.

Boulton, G. S. 1986. Push moraines and glacier contact fans in marine and terrestrial environments. *Sedimentology* 33:677–98.

Chivers R. C., N. Emerson, and D. Burns. 1990. New acoustic processing for underway surveying. *Hydrographic Journal* 42:8–17.

English Nature, Scottish National Heritage, Countryside Commission for Wales, Environment and Health Service (Department of Environment, Northern Ireland), Joint Nature Conservation Council, and Scottish Association for Marine Science. 2001. *Indications of good practice for establishing management schemes on European marine sites: Learning from the UK Marine SACs Project 1996–2001.* Peteborough: English Nature.

Foster-Smith, R. L., J. Davies, and I. Sotheran. 2000. Broad scale remote survey and mapping of sublittoral habitats and biota: technical report of the Broadscale Mapping Project. Scottish Natural Heritage Research, *Survey and Monitoring Report* No. 167.

Sotheran, I. S., R. L. Foster-Smith, and J. Davies. 1997. Mapping of marine benthic habitats using image processing techniques within a raster-based geographic information system. *Estuarine, Coastal and Shelf Science* 44, suppl. A:25–31.

Utilization of GIS in a Fisheries Assessment and Management System

Erik C. Franklin, Jerald S. Ault, and Steven G. Smith
Rosenstiel School of Marine and Atmospheric Science
University of Miami, Florida

Human population explosion, resource depletion, and habitat destruction; these common phrases are buzzwords for the changes that are occurring throughout our world. Nowhere are these transformations more evident than in coastal environments. The interface between land and sea has long enticed human cultures to enjoy the bounty of marine goods and services, yet only recently as our numbers surged have we begun to witness the consequences of our actions. Future reversal of these trends will depend upon policy from decision makers informed by sound scientific investigations. With this goal in mind, the Fisheries Ecosystem Modeling and Assessment Research (FEMAR) group at the Rosenstiel School of Marine and Atmospheric Science has developed and implemented a research program aimed at understanding marine ecological processes and translating that knowledge into useful guidelines for the adaptive management of marine resources. FEMAR utilizes a systems science approach to investigate landscape-scale marine ecological projects and fisheries assessments. Due to the spatial nature of the projects, GIS is used for sampling survey design, the delineation of habitats, analysis of habitat utilization by fish, and the evaluation of spatial fisheries management strategies.

GIS is an efficient technology to determine the spatial sampling framework for a fisheries assessment and management system *(figure 1)*. The system is an organized set of scientific protocols and methods designed to achieve three main goals: (1) to understand fisheries resources and habitats within the context of the aquatic ecosystem; (2) to assess the impacts of human activities and the source of economic demand on these resources; and (3) to analyze and evaluate the degree of success of proposed and implemented management policies in mitigating human impacts on fisheries resources.

Using GIS, an initial step in the system process is the creation of a grid overlying the research area *(figure 2)*. Each grid cell is provided a unique identity and includes attributes of the biotic and abiotic characteristics found within that cell. The grid cells with their associated attributes form the basic units for the planning of the sampling design. With this approach, GIS has proven most useful in visualizing the randomized sample locations and managing those locations for the consideration of future sampling allocations and optimal survey design.

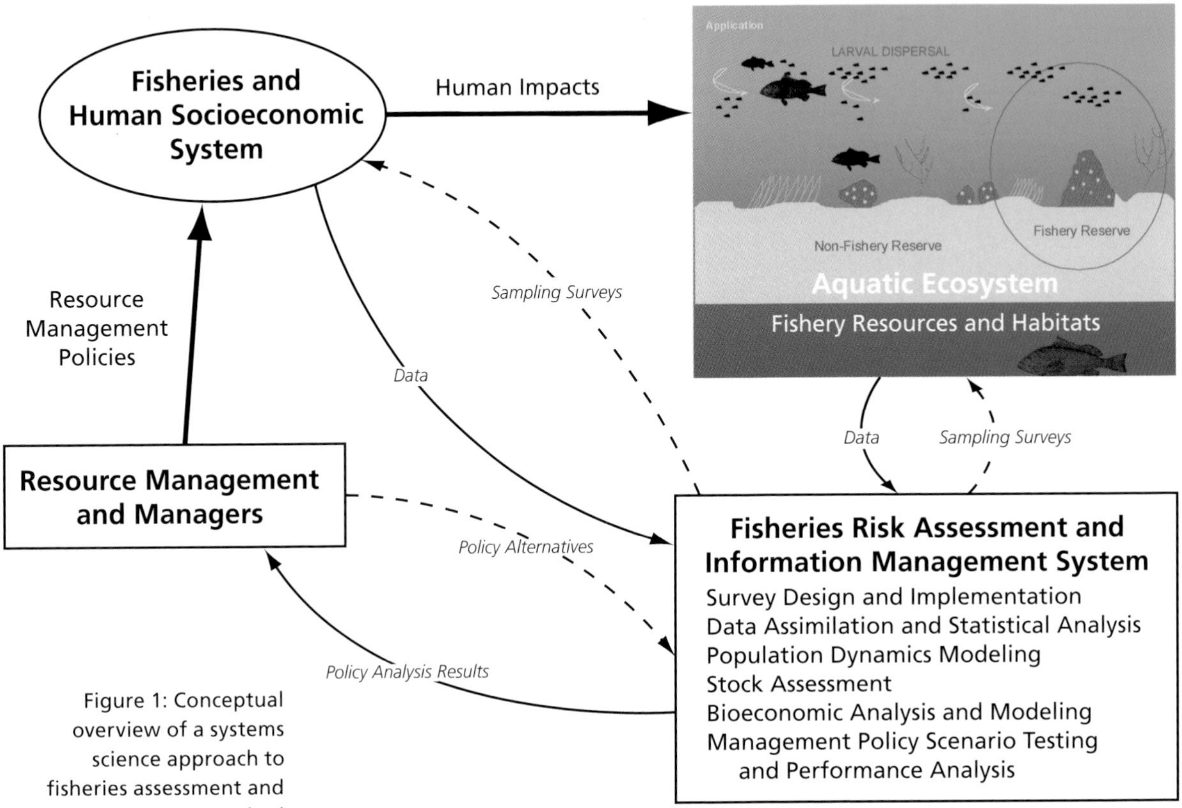

Figure 1: Conceptual overview of a systems science approach to fisheries assessment and management. Dashed lines indicate adaptive steps. Arrows indicate direction of impacts.

GIS provides an excellent tool to delineate the extent and association of marine habitats. In the Florida Keys, coastal bay habitats play a critical role in the function and dynamics of the coral reef ecosystem. These inshore waters serve as a nursery area for larvae and juveniles of a wide variety of fish and shellfish. Many of these animals live and reproduce in barrier coral reef and other offshore habitats as adults. The diversity and connectivity of marine habitats is directly linked to the capacity, status, and productivity of the associated fish communities. Visualization with a GIS allows a patchwork of different habitats to be linked in a spatial fashion that also relates to various ontogenetic stages for the fish. Coastal fish, particularly reef fish and shellfish, are ideal indicators of environmental stress (e.g., fishing, habitat changes, and so on) because they are tightly linked to the southern Florida coastal ecosystem. Observing and interpreting changes in the status of these indicator species requires a sampling strategy that integrates statistical survey design principles. This innovative process links digital maps of habitats such as benthic substrates, bathymetry, and coral reef benthic biota with underwater survey methodologies, and statistical associations between fish and habitats *(figure 3)*.

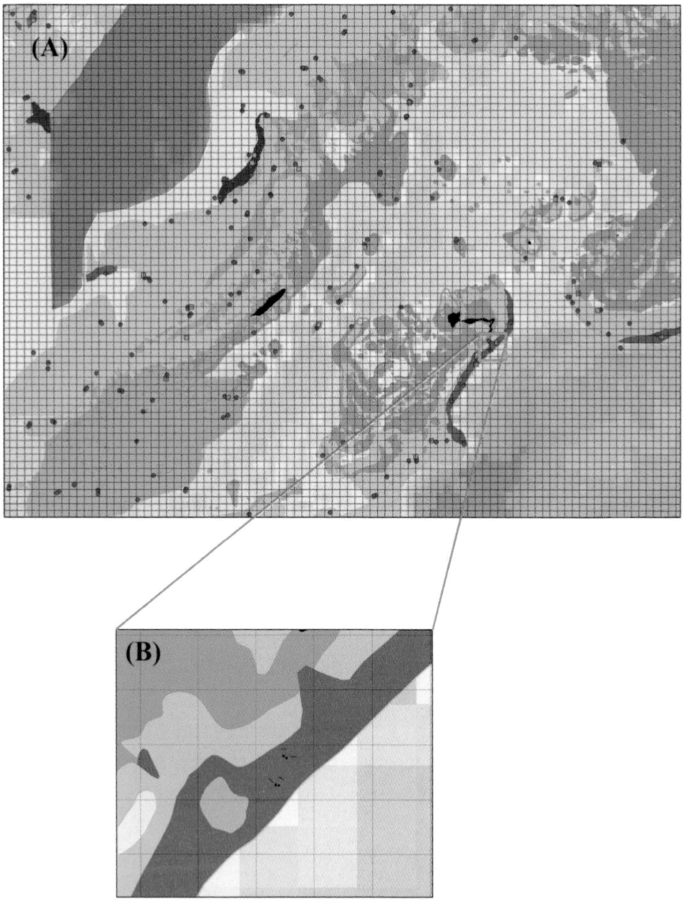

Figure 2: Benthic habitat map of a portion of the Florida Keys divided into a grid of sampling subunits used to determine the sampling allocation for the survey design.

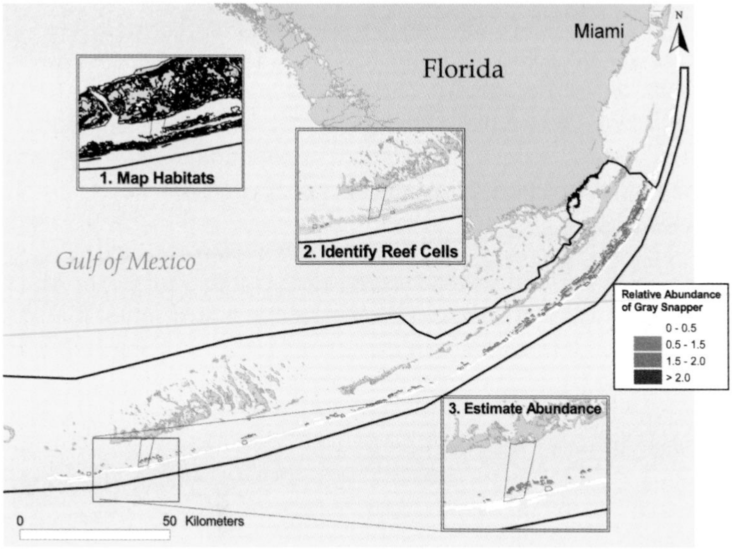

Figure 3: A visual representation of the relative abundance of gray snapper, a common reef fish species, that has been estimated for the extent of the coral reef ecosystem of the Florida Keys.

Figure 4: One of the FEMAR scientists tagging a bonefish caught for the acoustic telemetry project.

The movement patterns and habitat utilization of fish can be analyzed with GIS. Bonefish *(Albula vulpes)* and tarpon *(Megalops atlanticus)* in south Florida and the Florida Keys support internationally recognized and economically important recreational fisheries. Acoustic tracking allows analysis of fish movement patterns relative to the physical properties of the water, biological properties of the bottom habitats, and patterns of current movement, as well as climatic and seasonal changes in weather conditions *(figure 4)*. This technology can be used to examine movements over many different timescales. Daily movement patterns emerge from short-term tracking. Home range size and habitat selection will then be inferred from data gathered over days or weeks.

Billfish, a collective group of large migratory fish such as sailfish, marlin, and swordfish, inhabit coastal and pelagic oceanic waters of all tropical and subtropical seas. An important attribute of the billfish life history is that the stocks of these large pelagic fish display a strong population dynamic response to environmental variability. Similar to the large tunas, billfish experience a complex developmental process that occurs in a variety of habitat types. Using GIS, spatial models are being developed to identify "essential pelagic habitats" to predict the spatial distribution of the billfish. Stock migrations seem to be clearly linked to the seasonal patterns of sea surface temperatures. These seasonal migrations are evidenced in the spatial distribution of commercial catches.

Using GIS can assist in determining the efficacy of coastal zoning strategies. No-take marine reserves have recently emerged as a spatial management strategy to increase fishery yields while protecting marine biodiversity. The Florida Keys National Marine Sanctuary and the Dry Tortugas National Park have established a patchwork of no-take zones that encompass a variety of habitats, depths, and oceanographic regimes *(figure 5)*. The planning and placement of the marine reserves involved the use of GIS to identify target variables such as biodiversity, animal abundance, or unique habitats. Ongoing research to determine the impact of the spatial closures on fish and benthic communities actively utilizes GIS.

GIS is a powerful tool used to plan, implement, and analyze marine ecological projects and fisheries assessments. This project illustrates the diverse and useful ways in which a GIS can be integrated into a marine research and fisheries assessment program.

Additional resources

Fisheries Ecosystem Modeling and Assessment Research Group: *femar.rsmas.miami.edu*

Bonefish Research Project: *www.bonefishresearch.com*

Rosenstiel School of Marine and Atmospheric Science: *www.rsmas.miami.edu*

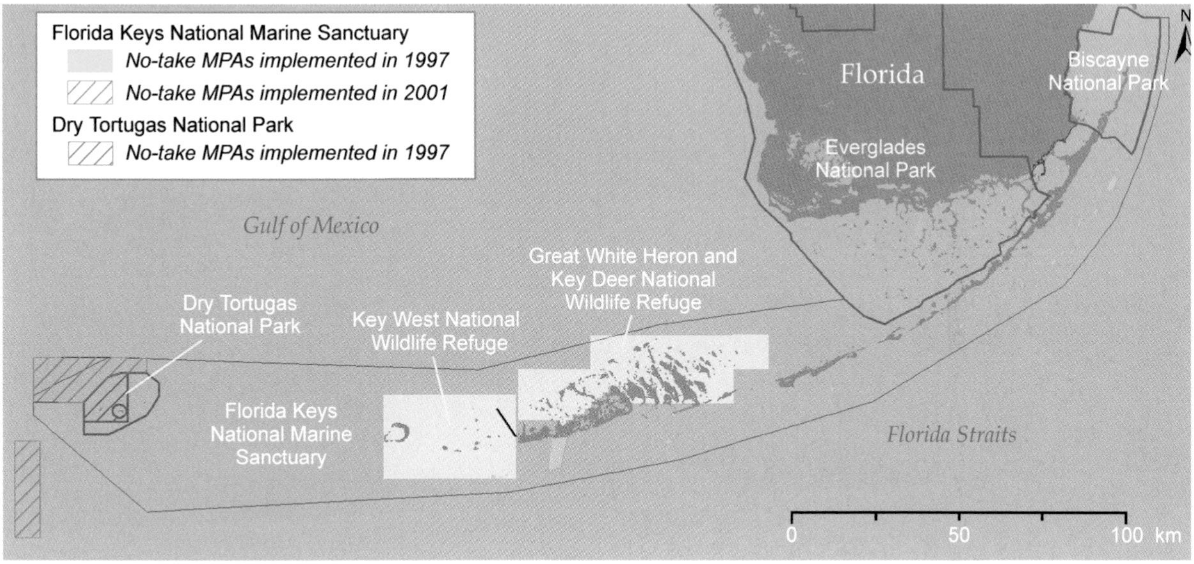

Figure 5: The zoning scheme of the Florida Keys National Marine Sanctuary illustrates the complexity of the spatial management strategies currently employed in the Florida Keys.

GIS Analysis of Landsat 7 Thermal Data to Identify Submarine Springs

James Alan Reade McClean
Florida Geological Survey
Tallahassee, Florida

The Florida Geological Survey's Coastal Research Group, a bureau within the Florida Department of Environmental Protection, has the responsibility of monitoring and evaluating statewide coastal geologic resources. The goal of this project, funded by the FGS and the U.S. Geological Survey, is to locate submarine groundwater discharge areas in Florida's coastal waters and to investigate their physical characteristics. Several freshwater springs have been reported for both the Gulf of Mexico and the Atlantic coast (Rosenau et al. 1977); however, scientific knowledge of these phenomena is sparse at present. Additional information of these important natural resources is required to monitor the effect of pollution from onshore sources, to understand the influences of groundwater discharge on fish habitats, and to identify potential sources of drinking water to meet the state's ever-growing population needs. Geographic information system (GIS) tools, including ArcView software from ESRI and ERDAS IMAGINE®, are used to analyze a variety of data sources to enable advance planning prior to expensive field investigations. Research vessel use is always limited by the availability of time and funds. GIS analysis of Landsat 7 imagery permits an inexpensive identification of specific target locations before setting out to sea. This economizes state assets by planning field investigations for locations of maximum potential interest.

Initial exploratory data analysis was performed on Landsat 7 satellite image data as a pilot study to evaluate the feasibility of using remote-sensing data to identify submarine groundwater discharge vents along Florida's coasts. Long-wave thermal data was chosen for this project with the hope of identifying groundwater discharge based upon a thermal signature (Bogle and Loy 1995). Florida's springs typically discharge groundwater at a relatively constant temperature throughout the year. Groundwater is less susceptible to seasonal temperature fluctuations that affect surface water such as ambient temperature, solar radiation, currents, and wind. Springwater in Florida maintains a relatively constant temperature of about 20 degrees Celsius throughout the year. In contrast, surface water in the Gulf of Mexico experiences dramatic seasonal variation, ranging from a high of 36 degrees Celsius in summer to a low of 15 degrees in winter.

The constant temperature of groundwater discharge and the seasonal fluctuations in seawater surface temperatures create two windows of maximal thermal difference: summer and winter. During late August and September, groundwater discharge should appear as cold spots in the otherwise warm Gulf of Mexico. A summer temperature variance of 8 degrees Celsius occurs between cooler groundwater discharge and the warmer Gulf of Mexico surface. The opposite occurs during January, when the Gulf of Mexico is at its seasonal lowest average temperature. Springwater at 20 degrees Celsius is a warm thermal anomaly when compared to the Gulf temperatures in the mid-teens. During winter, this warmer, less dense water rises to the surface to form a layer over surrounding saltwater zones.

Landsat 7 satellite imagery of the study area was provided by the Florida Department of Environmental Protection (FDEP). Scenes were available from December 1999 to January 2000. This time of year is typically dry, yielding cloud-free imagery more consistently than in summer months, when heavy rains prevail. It is also assumed that warmer, less dense freshwater will rise and is more easily detected in winter than in summer, when warmer Gulf of Mexico waters are more likely to mask freshwater upwelling. Data imagery utilized for this study included three scenes covering areas of primary interest to this project: P18R39, P17R40, and P16R39. These scenes are labeled using the WORLDREF Path/Row naming convention. Scene path numbers increase from east to west while row numbers increase from south to north. This data was georectified by FDEP personnel to a custom Albers Conformal projection. This is a standard FDEP projection for geographic data products and is consistent with several other available data sets used for spatial analysis in a GIS

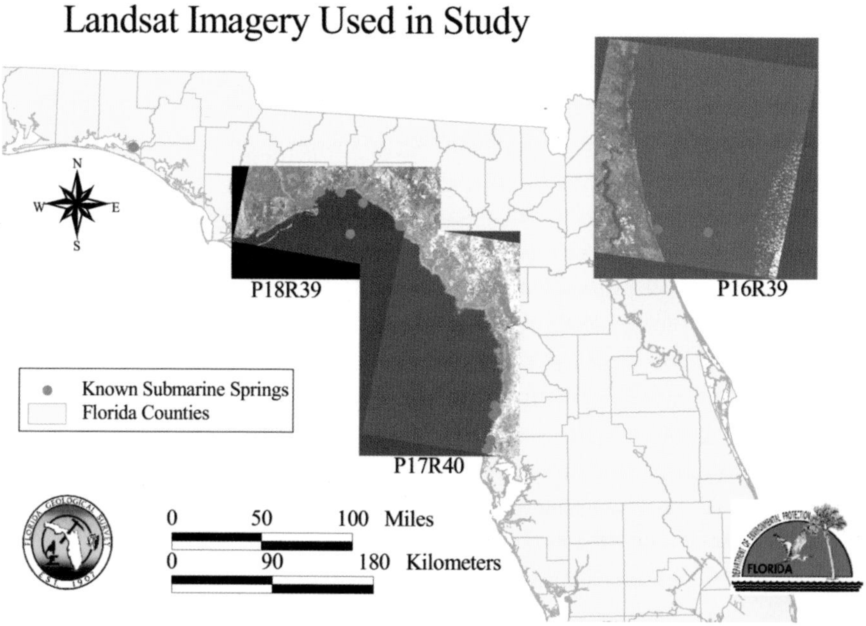

Figure 1: Landsat 7 ETM+ Band 6 imagery used for this project. Circles indicate location of previously known submarine freshwater discharge springs after Rosenau et al. 1977.

software package such as ArcView from ESRI. The three scenes utilized for this study were selected because they correspond to areas that have known reported submarine groundwater discharge springs. The scenes used cover the northeastern Gulf of Mexico known as the Big Bend Region, southward along the west coast of Florida to Clearwater Beach, and a portion of Florida's east coast along the Atlantic Ocean from the Georgia border south to Volusia County. Figure 1 illustrates the available Landsat imagery with plotted locations of known submarine springs(location of springs from Rousenau et al. 1977).

Data analysis

Landsat 7 carries on board the Enhanced Thematic Mapper Plus (ETM+) sensor. This instrument collects data in eight spectral bands. Seven of the bands utilize the visible and near-infrared spectrum and have a spatial resolution of 28.5 meters on the ground. Band 6 utilizes thermal frequencies. These longer wavelengths result in decreased spatial resolution due to decreased sensor head sensitivity. Band 6 has an effective ground spatial resolution of 57 meters, meaning each pixel covers 57 square meters in area along the earth's surface (Campbell 1996). Previously identified submarine springs are generally less than 57 square meters in extent. Therefore, the thermal band data has coarser spatial resolution than the features of interest. First-magnitude springs, like Wakulla Spring flowing overland south of Tallahassee at greater than 693 million liters per day, have recognizably large discharge areas (Lane 1986). One aim of this project is to identify offshore submarine analogues to Wakulla Spring. A basic assumption of this analysis is that less dense freshwater rises and expands over colder seawater. For smaller submarine springs, lenses of lighter warmer water are predicted to be larger than the vent features themselves. This means that winter data should allow for detection of smaller-size discharge springs than possible in summer. Radiometric sensitivity of Band 6 thermal data from Landsat 7 is 2 degrees Celsius. This implies that temperature differences greater than 2 degrees are detectable using this analytic procedure (Jensen 2000).

Thermal data analysis

Image classification was utilized to identify thermal anomalies. Supervised and unsupervised thematic map classifications were the two methods employed. This was performed in ERDAS IMAGINE 8.4. These are basic tools for classifying remotely sensed data into general user-defined categories (Lillesand and Kiefer 2000). Unsupervised classification entails statistical analysis of the pixel value range in a scene and then assigning these pixels to a user-defined number of classes. Pixel values were assigned to a class based upon their maximum likelihood of belonging to a particular class. Only Band 6 was utilized for this classification because it contains the thermal properties of interest to this study. Pixel values in the original image scene range from zero to 255. After the classification algorithm is performed, a new image is created that assigns each of the original pixels to one of 12 classes. Because only Band 6 thermal data is utilized, the new classes range from coldest to warmest in value, class 1 being coolest and class 12 being warmest.

To correctly label the resulting classes, it is necessary to assign names based upon operator knowledge of the terrain under investigation. This can be accomplished by reference to ground observations, high-resolution aerial imagery, or user knowledge of the area. Although local areas were accessible for ground truth observations, this project made extensive use of high-quality, 1-meter resolution digital ortho quad (DOQ) aerial images. These images were also available from FDEP in the Albers custom projection. This coverage is available for the entire state of Florida. Comparison of the DOQ image to the 12-class thematic image enabled the assignment of class names. The warmest classes are clearly related to barren lands, urban areas, and sandy beaches. The classes range from cooler objects such as various types of vegetation, down to the coolest values, which represent water bodies. Four classes are defined that represent water bodies.

Because water absorbs nearly all incoming incident thermal radiation, only background emitted radiation for water is detected in Band 6 (Jensen 2000). A narrow range of pixel values represents water in the original Band 6 images. These pixel color values range from 7 to 13 on a scale of zero to 255. This represents about 2.3 percent of the total possible range of values. The narrow range of thermal values detected over water bodies was sufficient for this pilot study. An alternative solution is to perform aerial thermal scanning missions over selected target areas. Data of this type would have a wider temperature sensitivity range and better ground spatial resolution. However, airborne thermal imagery missions are more expensive than readily available spatial data. This GIS data-

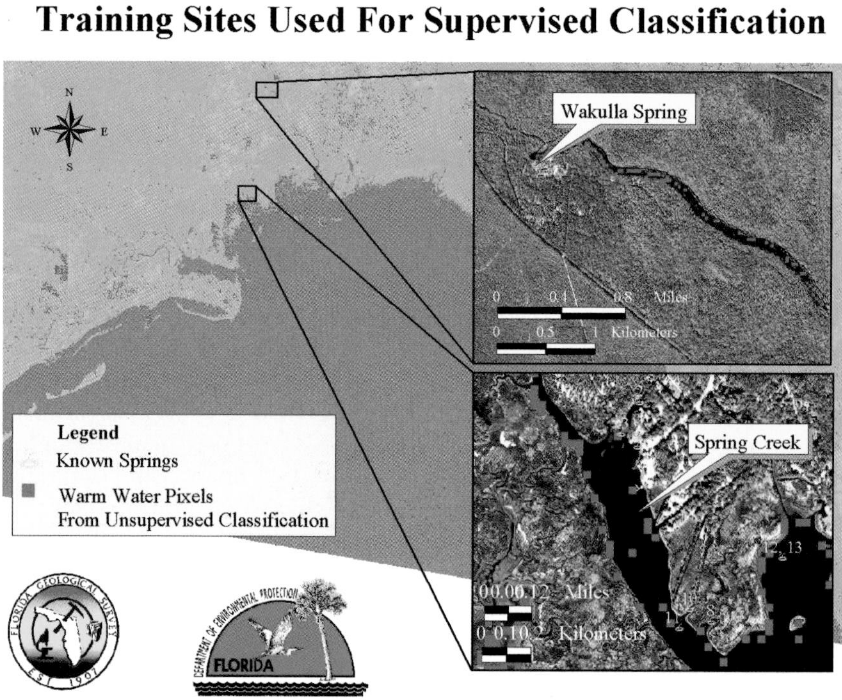

Figure 2: Test areas used for supervised classification of groundwater discharge. Upper right shows Wakulla River emanating from Wakulla Spring. Lower right shows several spring features at Spring Creek. Red pixels represent warmer water pixels. At this stage of analysis it is difficult to distinguish spring discharge from sandbars.

analysis pilot project resulted in specific target areas for future custom surveys. This pilot study helps reduce the high cost of aerial reconnaissance missions by narrowing search locations to clearly defined study areas. This will result in lower-altitude flights, smaller-scale surveys, and better image resolution.

The second form of image classification utilized in this project is supervised classification. Data from known areas is used to define classes (Lillesand and Kiefer 2000). The ERDAS IMAGINE operator defines classes in a signature editor file. A new image is created based upon these user-defined classes. Pixel values are utilized for this project to define freshwater-flow thermal characteristics over known groundwater springs. Pixels over the Spring Creek study area and Wakulla Spring and ensuing Wakulla River were defined as spring and river values *(figure 2)*. Sandy beaches and vegetation areas were assigned their respective classes as identified in the 1-meter-resolution aerial photos. Once defined, this signature editor file is used to create new image classifications. Both the supervised and unsupervised image classifications had nearly the same results. These two scenes are then statistically compared to check for overlapping pixels of the same values. This is done in ERDAS IMAGINE as an accuracy assessment using a matrix of both 12-class images or in ArcView Spatial Analyst to define areas of similar value in the two map layers.

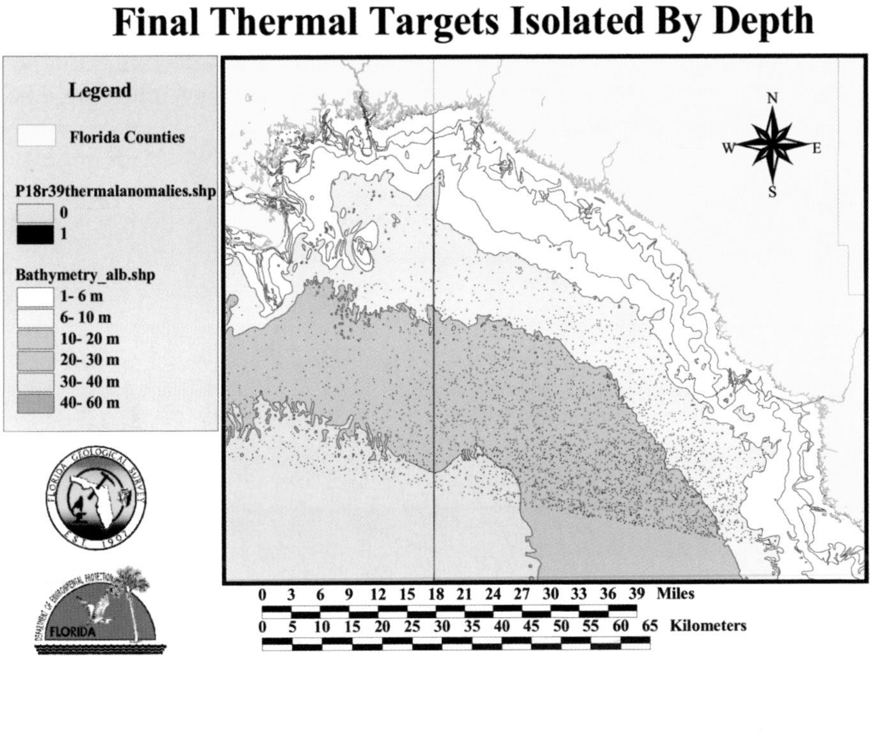

Figure 3: Subset of thermal anomalies selected from the union with water depths between 6 and 60 meters.

ArcView software was also utilized to analyze a variety of input data. A database of known wet sinks and flowing offshore springs was created and utilized as a base data set to explore spatial clusters of the phenomena under investigation. Bathymetry data was utilized to help differentiate warm thermal anomalies occurring over shallow water areas from possible offshore spring targets discharging in deeper water. Since all known offshore springs were recorded as being in water less than 60 meters deep, this depth was used as a threshold on which to base the selection of potential targets. Overlaying ArcView shapefiles of suspected freshwater discharge locations (i.e., thermal anomalies) over bathymetry data and selecting only targets in water between 6 and 60 meters deep was used to isolate final target areas. Results of this are presented in figure 3. A list of potential targets was created from the subset of these joined shapefile data sets. Locations having two or more contiguous target pixels were selected from the resulting data set. This was to isolate targets of larger size that would be easier to locate during field investigations.

Conclusion

This GIS analysis was conducted using remote-sensing imagery and conventional computer cartography data products available to our agency at no cost. Operator expense was the principal financial requirement for this stage of the project. The Florida Geological Survey possesses a variety of marine survey technologies and a fleet of capable research vessels. However, mobilization of these field assets requires a commitment of funds, personnel, and time. GIS analysis enables investigators to better plan and organize field operations by narrowing down potential survey areas. GIS software is also used to plan survey patterns and calculate time and fuel budgets in advance. This pilot project has successfully demonstrated the capabilities of GIS analysis in reducing research expenses by maximizing efficient deployment of field survey teams.

References

Bogle, F. R., and K. Loy. 1995. The application of thermal infrared photography in the identification of submerged springs in Chickamaugua Reservoir, Hamilton County, Tennessee. Paper presented at the Karst Geohazards, Engineering and Environmental Problems in Karst Terrain. *Proceedings of the Fifth Multidisciplinary Conference on Sinkholes and the Engineering and Environmental Impacts of Karst, Gatlinburg, Tennessee*, 415–24.

Campbell, J. B. 1996. *Introduction to remote sensing.* New York: Guilford Press.

Jensen, J. R. 2000. *Remote sensing of the environment: An earth resource perspective.* Upper Saddle River: Prentice Hall.

Lane, E. 1986. Karst in Florida: Florida Geological Survey Special Publication No. 29, Florida Geological Survey, Tallahassee.

Lillesand, T. M., and R. W. Kiefer. 2000. *Remote sensing and image interpretation. 4th ed.* New York: John Wiley & Sons.

Rosenau, J. C., G. L. Faulkner, C. W. Hendry, Jr., and R. W. Hull. 1977. Springs of Florida. *Florida Bureau of Geology Bulletin* 31 (rev.), Tallahassee.

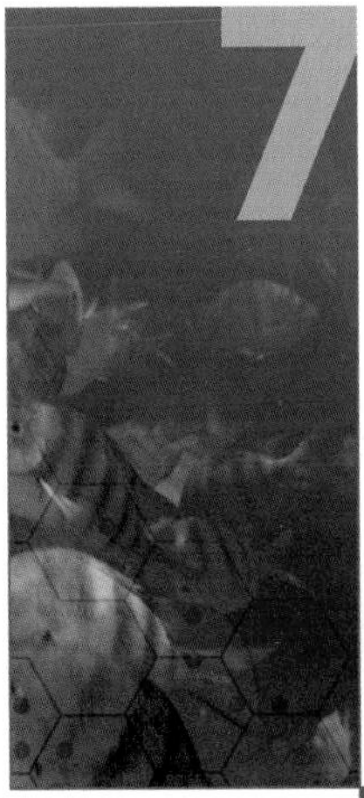

7 Monitoring Dolphin Behavior and the Effects of Restoration

Laura K. Engleby
Dolphin Ecology Project
Key Largo, Florida

The ecological degradation of South Florida is as notorious as ecological degradations in Lake Erie and Chesapeake Bay. Humans have altered the South Florida landscape in ways that affect the temporal and spatial variability in water flow, nutrient loading, and productivity. Florida Bay in particular has experienced increased salinity due to the diversion of freshwater input. This increased salinity, together with elevated nutrient levels from land development sources, has stimulated algae blooms, resulting in large-scale die-offs of sea grasses, sponges, and mangroves. Consequently, declines in these habitats have caused reductions in fish populations.

Over the past 50 years, this severe degradation has been well documented by scientists. During this time, top predators such as herons, brown pelicans, alligators, and storks have declined by 80–95 percent. Sixty-eight species of South Florida's mammals, birds, reptiles, amphibians, and plants are threatened or endangered. Federal and state management agencies have responded to these declines with an intensive, $7.8-billion restoration project designed to restore the natural quantity, quality, timing, and distribution of freshwater into Florida Bay. No one knows what impacts these changes will have on a highly visible, upper trophic-level species—the bottlenose dolphin.

Regional managers of federal, state, and county agencies need information. Currently, agency managers overseeing the restoration efforts have no baseline information about bottlenose dolphin numbers, preferred habitats, seasonal movement, food requirements, or reproduction. Restoration projects require essential information on life history, density, and distribution patterns of bottlenose dolphins as well as how they relate to, and are affected by, habitat and water

During the habitat use study, researchers observed "mud-ring feeding," a previously undescribed feeding strategy by bottlenose dolphins in Florida Bay. This feeding strategy is closely associated with the mud banks in the bay. Feeding either takes place along the edge of the bank or on the bank itself.

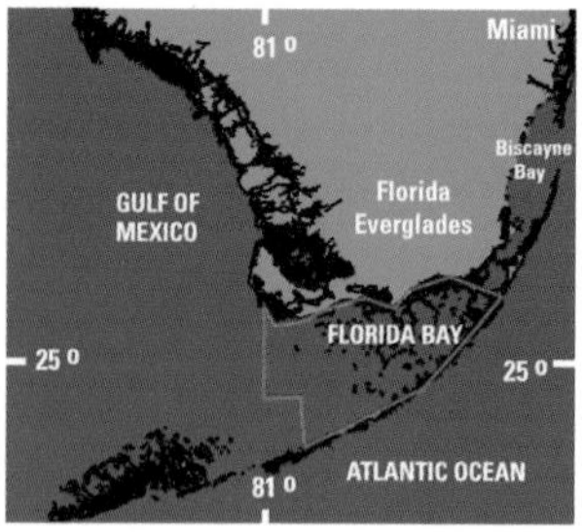

Using researchers, volunteers, recording equipment, and a GIS, the Dolphin Ecology Project hopes to collect and distribute information to restoration agencies to assist in decision making.

quality in the Florida Keys. Using researchers, volunteers, recording equipment, and a GIS, the Dolphin Ecology Project (DEP) hopes to collect and distribute this information to restoration agencies to assist in decision making.

The project's GIS is useful in determining:

- The distribution and density of dolphins that regularly live around the Florida Keys and the number of dolphins that might be transiting through this area.
- The types of prey that dolphins depend on and the types of habitats that dolphins utilize to find their prey.
- The impact on dolphins produced by changes in their environment such as variations in salinity, nutrient levels, and pesticide runoff.
- The distribution of bottlenose dolphins and their prey in relation to water quality. Surveys conducted throughout the year reveal seasonal changes in both dolphin distribution in the area as well as dolphin habitat use. To do this, the team divides the study area into strata representing habitats of similar composition and quality. Moreover, they assess prey distribution using a small otter trawl.

The team also measures other properties of habitat qualities such as salinity, temperature, and turbidity.

GIS is instrumental in recording dolphin distribution and feeding behavior. When the team locates a pod of dolphins, they record the sighting location using a differential Global Positioning System (GPS) unit. They also take photographs of the dorsal fin of each dolphin in the group to recognize individual dolphins. One distinctive dolphin is selected as a focal animal and is then followed from a distance of less than 100 meters. During each follow, the focal animal's location, habitat type, behavior, group composition, and size are recorded at three-minute intervals.

Researchers record the behavioral state of the dolphin using the following criteria:

- Travel—directed movement in a specific direction.
- Socializing—tactile contact among group members.
- Resting—slow, quiescent movement or individuals remaining motionless at the surface.
- Feeding—direct evidence (prey in mouth) or strong indication of feeding (chasing prey, rapid changes of direction).

Using these criteria, the observers are able to distinguish dolphins that are actively feeding from those engaged in other activities. Once a dolphin is seen feeding, researchers then sample prey (using a small otter trawl) and water quality (temperature, salinity) using a YSI multiprobe. Thus, they obtain samples of fish abundance, species composition, and water quality at sites where dolphins are feeding. During the follow, researchers sample prey and water quality at regular intervals to obtain control data. This distribution, movement, and

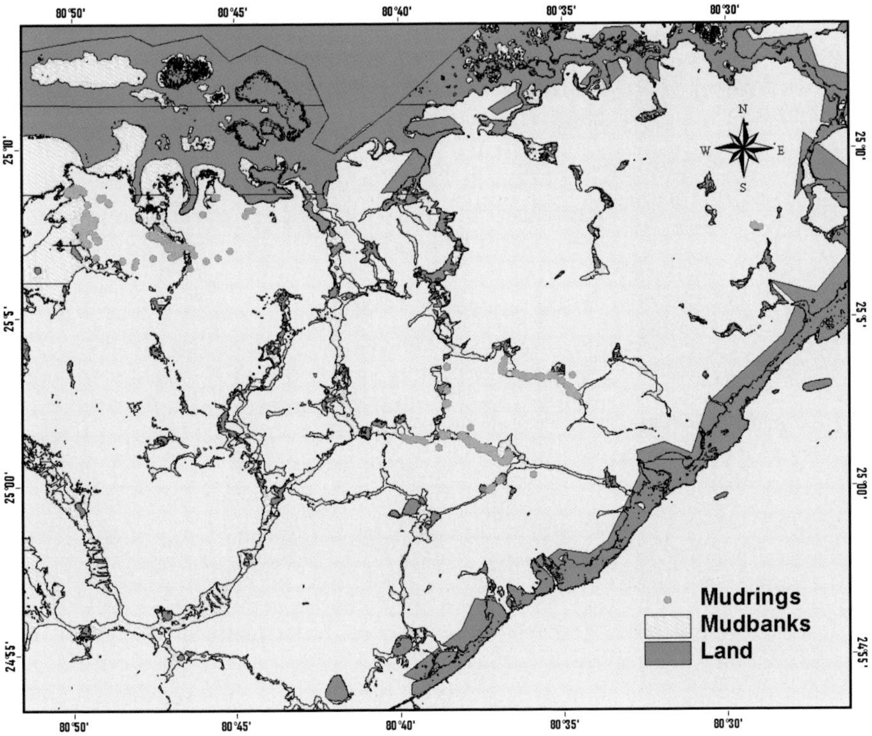

Central and eastern Florida Bay mud-ring feeding sites.

behavioral data collected in the field is then incorporated into ArcView to test the hypothesis that dolphins feed preferentially in habitats where water quality is good and prey densities are high.

In addition to these dedicated focal-animal-follows, the DEP examines data on dolphin distribution collected by its collaborators at the National Marine Fisheries Service South East Fisheries Science Center. Eventually, the DEP would like to overlay distributional information from the U.S. Coast Guard aerial surveys to see which habitats dolphins are using on a large scale and how their use of the habitat varies seasonally. This collaboration will disclose patterns of dolphin habitat use over a variety of spatial scales.

GIS-generated maps depict the current distribution of dolphins and prey sampling sites both for control setup and in the presence of feeding/nonfeeding dolphins. Also, maps will be generated that depict the distribution of dolphins in comparison with benthic communities and water quality parameters.

By employing a geographic framework for evaluating linkages between habitat quality, prey densities, and dolphin distribution, the DEP can provide sound, scientific advice on the effects of restoration options on the population(s) of bottlenose dolphins in Florida Bay and in the Florida Keys. The GIS generates a comprehensive picture of the relationship between ecosystem health and dynamics of habitat use by bottlenose dolphins. This information will be valuable to managers attempting to predict, monitor, and understand the effects of ecosystem restoration efforts in South Florida.

The Dolphin Ecology Project has conducted its pilot field program, which was funded by the National Marine Fisheries Service, in Florida Bay. Although the analysis is not complete, some preliminary assessment of the distribution and behavior of bottlenose dolphins in Florida Bay has been made. Much analytical work remains to be completed, and any conclusion presented here should be regarded as provisional pending further analysis.

The surveys from the project confirm that dolphins are present in Florida Bay throughout the year. The photo identification catalog presently includes 108 identifiable individuals for the eastern Florida Bay study area. A qualitative examination of the distribution of dolphin sightings and focal follows reveals that most dolphins were found in the southern portion of the study area, where the environment is influenced by exchange with the Atlantic Ocean.

The analysis of the spatial variation of potential prey is not yet complete, but other studies have found that fish densities are relatively low in the northern Florida Bay. It is believed that the distribution of bottlenose dolphins in this area is primarily driven by the distribution of prey, which, in turn, is determined by environmental factors such as temperature, salinity, and habitat type. Due to their high visibility, bottlenose dolphins are a good indicator of the distribution of prey and, in turn, the quality of habitat and environment that support these animals.

The South Florida Ecosystem Restoration Program is the largest environmental restoration project ever attempted in the United States. Along with other critical projects that are under way in South Florida, its goal is to restore the regional hydropattern and recover endangered species and habitats. Within this affiliation, the Dolphin Ecology Project has an unprecedented opportunity to establish baseline information for bottlenose dolphins in Florida Bay before a major environmental restoration project takes place. Researchers can then evaluate the effects of restoration efforts on a top predator in this ecosystem.

The Dolphin Ecology Project outreach work consists of participation in scientific meetings and direct interaction with managers and stakeholders interested in bottlenose dolphins and their habitats in South Florida. At the conclusion of the field project, the team intends to hold several public meetings in South Florida in order to highlight its findings and promote restoration programs using the bottlenose dolphin as a flagship species for these efforts.

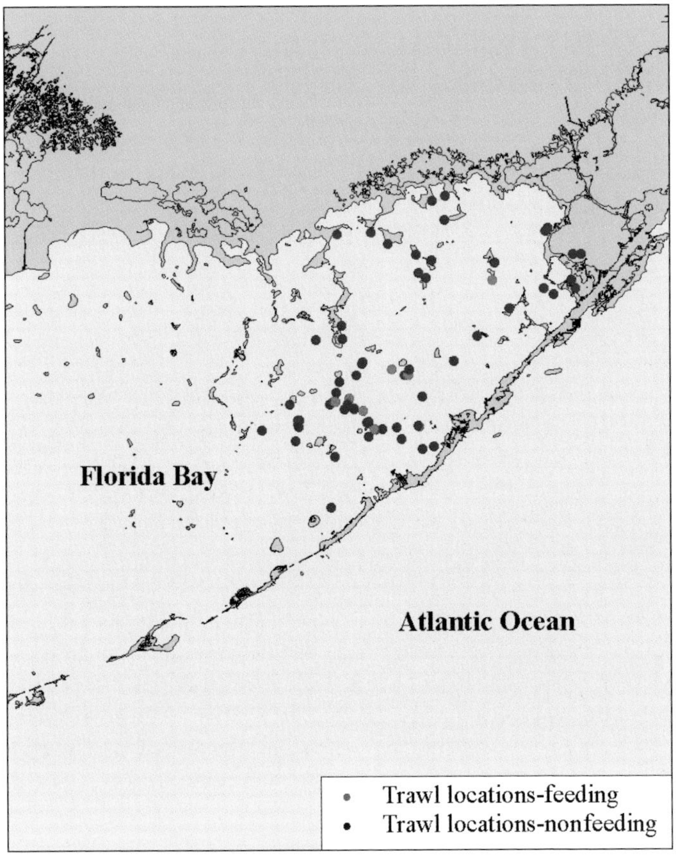

PHOTO COURTESY OF MICHELLE KINZEL

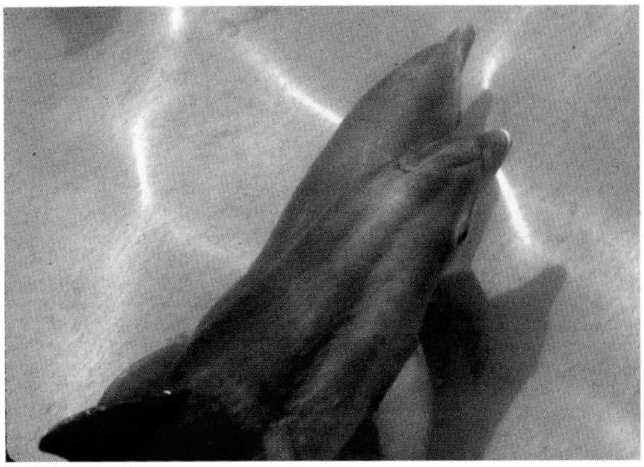

The work described here represents some of the first dedicated research on bottlenose dolphins in Florida Bay. The project team believes that continuation of this research will provide a rich baseline of information for assessing the response of bottlenose dolphins to future changes in the south Florida ecosystem. An increase in the density and range of bottlenose dolphins would be clear evidence of the benefits of restoration programs.

In general, public support is essential to the success of any management program. The Dolphin Ecology Project study of the bottlenose dolphins will surely be an effective tool for educating the public about the complexities of ecosystem protection and restoration.

Over the past 50 years, severe environmental degradation has been well documented by scientists. During this time, top predators such as herons, brown pelicans, alligators, and storks have declined by 80–95 percent.

Spatial Correlation Between Bluefin Tuna and Sea Surface Temperature Fronts

New England Aquarium

Robert S. Schick
Edgerton Research Laboratory, New England Aquarium
Boston, Massachusetts

At the New England Aquarium's Edgerton Research Laboratory, one research priority is to analyze the distribution and abundance of pelagic vertebrates, including several species of marine mammals as well as bluefin tuna. We are interested in using GIS to track the movements and distributions of right whales and bluefin tuna in the Gulf of Maine. This research has led to an analysis of the relationship between whales and tuna and the thermal structure of the ocean surface. While oceanographers will be quick to point out the necessity of exploring the z-dimension (depth measurements), our distribution data is primarily recorded at the surface. Thus we could explore the spatial correlation of right whales and tuna in relation to physical features. By using the Distributed Oceanographic Data System (DODS), we access a wealth of oceanographic data, most notably the Advanced Very High Resolution Radiometers (AVHRR) database from the University of Rhode Island, used widely to sample sea surface temperature. Using DODS, Perl, and ArcInfo, we set up a comprehensive GIS to quantify the spatial relationship between species distribution and sea surface temperature (SST) gradients, or fronts. The goal of this work is to see if we can use remotely sensed environmental variables to help explain the observed distribution of these species.

We have outlined the focus of our research and included many examples of the types of research questions we are exploring. Species–environment interactions can be attributed to distribution in relation to physical features (depth, slope, temperature, temperature gradient, and so forth) and can be influenced by the presence of conservation features, such as marine protected area (MPA) boundaries. Analyzing the spatial interaction between bluefin tuna and SST fronts, our goals for the research were threefold:

- To document the habitat-use patterns of right whales and bluefin tuna.
- To synthesize data from a variety of research platforms.
- To use spatial analysis to quantify the species–environment relationship.

In this chapter, new images, preliminary results from the spatial analysis, and ideas about future directions for the research contribute to the presentation of this GIS-based research.

Methods

Data from a variety of sources, including aerial survey, line-transect, and satellite tags, are used in the conversion to ArcInfo coverages and raster grids. This data loaded into a GIS allows for easy visualization, and prompts new research questions and directions. Once the species distribution data was intact, building the environmental database included outlining the initial steps to get the SST frontal data into ArcInfo. (For details on this, visit our Web site at *www.marinegis.org*). We were also interested in the relationship between species and SST. To explore this relationship, data manipulation between the university's AVHRR database, Matlab, and ArcInfo was conducted. The GIS work has been developed providing a mechanism to store, view, and manipulate both species and environmental data.

Once the databases were complete, a series of ARC Macro Language (AML™) scripts were developed to extract the values of each of the five environmental variables at specific dates and times. The initial focus of this database was on a number of variables: SST, gradient, density, slope and depth, and exploring the relationship between species and ocean color, as well as prey distribution. We considered using SeaWiFS (Sea-Viewing Wide Field Sensor) data collected as a part of NASA's Mission to Planet Earth; however, this data does not temporally overlap the initial species data sets. Some prey data is being made available from the National Marine Fisheries Service's (NMFS) Northeast Fisheries Science Center in Woods Hole, Massachusetts, and will be incorporated into the GIS. These values were formatted and imported into S-plus for statistical analysis (see Schick 2001a and Kraus et al. *in prep* for detailed explanations of the application of these techniques). Obtaining statistical correlations is fairly straightforward, while the harder and more complex task is interpreting the meaning of the results.

Results

Right whales

The analysis of the satellite tracks focused initially on two right whales in the Gulf of Maine (Schick 2001a, Kraus et al. *in prep*). One of the whales, a mature female with a calf, moved around the Gulf of Maine, heading north out of the Cape Cod region, and then turned south when she encountered colder water *(figure 1)*. We were interested in seeing if the patterns of right whale movement changed in relation to changes in the environmental features. To do this we ran a point-density filter over the whale track to create an index of point density such that locations where the whale ceased traveling had higher point density than locations where the whale was traveling. The analysis of this data focused on the relationship and similarities between the areas of highest point density. Our assumption was that slowly moving whales were more likely to be feeding and would therefore be found closer to SST fronts. Our attempts to examine

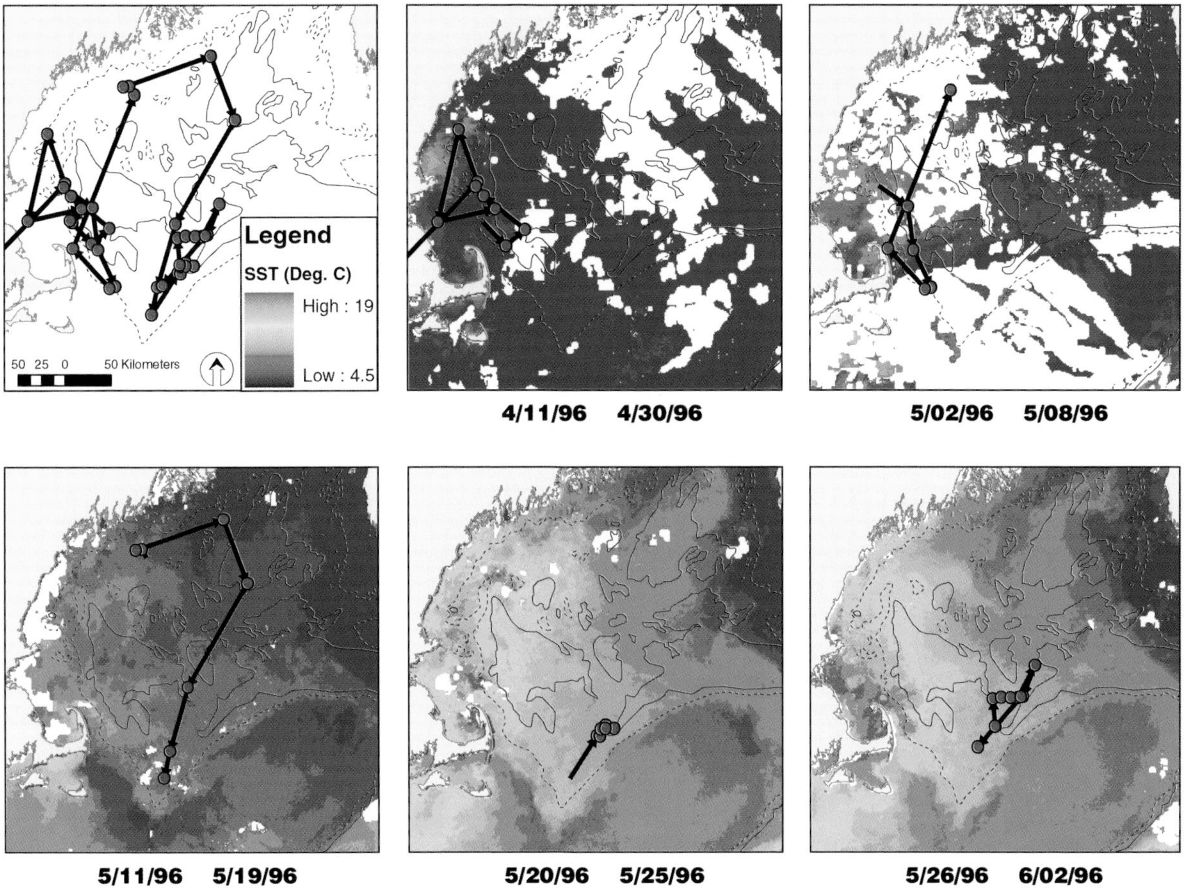

Figure 1: Gulf of Maine portion of one right whale mother and calf pair's track with accompanying AVHRR SST image. The track is broken into portions to provide a glimpse of the whales' movement in response to shifting temperature patterns in the Gulf of Maine. Each panel includes a three-day composite SST image.

the paths of the whales in relation to SST and SST fronts have indicated that the whales respond differently to their environment. Both the Mantel tests and the CART analysis indicate that temperature is a driving factor for the mother and calf pair.

Bluefin tuna

The interpretation of the bluefin tuna study is still ongoing, and some patterns have emerged. There was significant correlation of these dynamic predators *(figure 2)* relating to one variable at different times in the season; however, the results were sporadic. Tuna seemed to aggregate near fronts if there was a strong thermal gradient *(figure 3)*, most likely in response to the presence of prey. However, tuna would often abandon an area even with the presence of a strong thermal gradient, suggesting that perhaps the presence of prey species is more important than a persistent gradient.

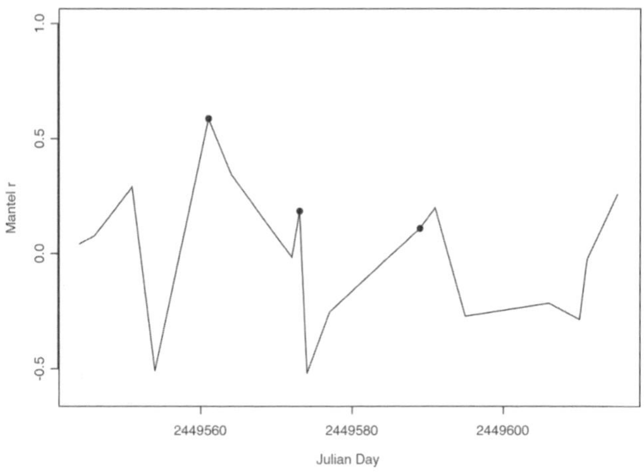

Figure 2: Graph statistical correlation between tuna presence and distance to front variable in 1994. Solid circles denote significant correlations between tuna and fronts. The first significant event corresponds to the middle panels in figure 3 (8/8/94).

Discussion

One key difference between marine and terrestrial systems is that the marine systems are not static. Certain variables like depth and slope remain relatively static, while the fluid nature of the ocean compounds the difficulty of associating mobile predators with their environment. In addition to the difficulties associated with data collection at sea, there are numerous difficulties in the study of a dynamic predator. For example, tuna do exhibit spatial correlations with fronts over one season, yet this was not consistent *(figure 2).* Tuna are able to range over a large extent quickly, and while they do show some significant associations with fronts, it is likely that they are responding to other features as well. Two of their preferred prey species, herring and mackerel, are themselves pelagic predators, and move in response to prey distribution. Adding prey data will help us explain observed distribution patterns that are left unexplained by our existing model. The tuna's response to a dynamic thermal event in 1994 seems to confirm that the movement may be toward the tracking of prey as opposed to being a result of physical features. As can be seen in the progression of images, tuna abandoned an area even when fronts seemed to persist, suggesting that while the large frontal event may be an important mechanism for prey aggregation, once the prey resources are exhausted, bluefin will leave in search of other feeding opportunities.

The analysis has generated a few interesting questions. Notably, we are extremely interested in the role that prey distribution plays in patterning bluefin tuna distribution. The addition of the prey data from NMFS will help us fill in that story, as will existing research in our lab that is focusing on digestive content analysis of bluefin in the Gulf of Maine. In addition, we noticed a large shift in distribution of tuna from 1994 to 1996 (Schick et al. *in prep*). As part of that shift we saw a great number of schools in the Stellwagen Bank region of the Gulf of Maine. Using a high-resolution raster bathymetry layer, we are exploring how and why tuna populate this region *(figure 4).* Initial exploration of this data suggests that depth is important at smaller scales than were tested in the above Mantel-based analysis. The entire set of Mantel-based results indicates that it is hard to make blanket statements about how bluefin use the Gulf of Maine over such a long period of time, and that we may have to parse the data into even smaller spatial subsets to help tell the ecological story.

The right whale results present an interesting result. While we have tested only a few whale tracks, we have used a powerful suite of statistical analyses. The results have been consistent across different tests, and suggest that whales indeed respond to the presence of fronts differently depending on their life history (Kraus et al. *in prep*). This result is intriguing because it suggests that while we can use fronts as a predictor variable in our research efforts to map out critical right whale habitat, (1) we may have to adjust our basic assumptions

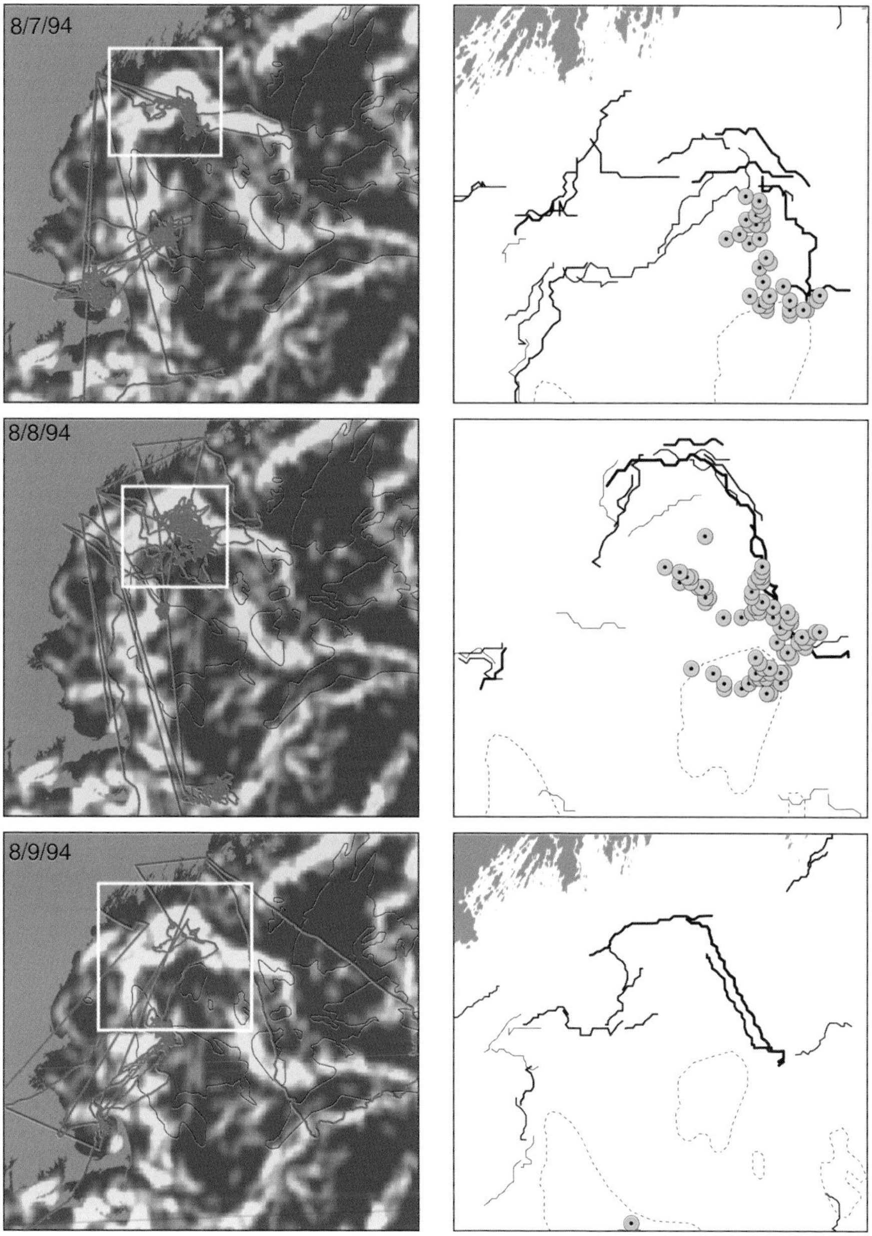

Figure 3: Bluefin tuna distribution patterns for three days in early August 1994 display frontal density and movement (left column), and school locations and SST front locations (right column). In the left column, all tracks of individual spotter planes (dark gray lines) are drawn over an index of SST frontal density (red values correspond to high frontal density). In the right column, locations (orange circles) are noted in relation to the actual thermal gradient (red lines). Fronts are drawn using graduated symbols, with thicker lines representing stronger gradients.

about the association with fronts, and (2) we need to create a different model for different life-history classes. Though interesting, we have to caution against drawing too much inference from such a small data set. Further analyses of this type may lead us to new understandings of right whale distributions, and may increase our understanding of the habitat requirements of this critically endangered species.

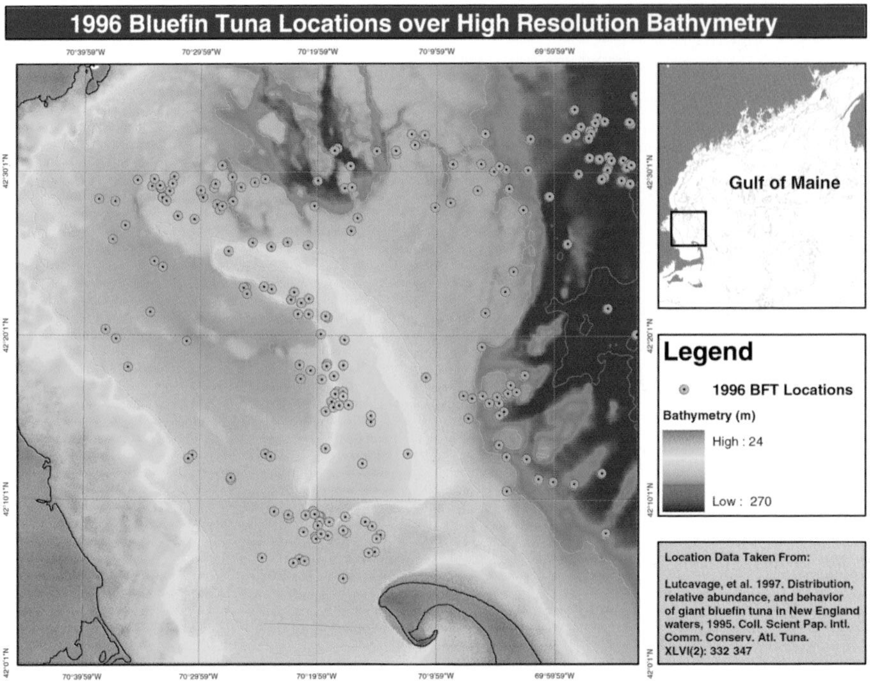

Figure 4: Bluefin tuna locations in and around the Stellwagen Bank area as seen in 1996. The depth layer has a 1-meter resolution. Tuna seem to prefer the western edge of the bank, and do not seem to congregate over the shallower parts of the bank.

The research we have conducted so far at the New England Aquarium has led to some exciting results, notably that both right whales and bluefin tuna movement does exhibit significant correlation with sea surface temperature gradients. However, the response for each species is highly dynamic. Therefore we are looking to additional environmental data sets to help flesh out our understanding of how these species utilize their environment. Ultimately such an understanding will yield important information for researchers and critical information for conservation purposes.

References

Kraus, S. D., R. S. Schick, and C. S. Slay. *Behavioral and environmental analysis of satellite tagged right whale movement tracks.* In preparation.

Schick, R. S. 2001a. Tuna distribution in relation to physical features in the Gulf of Maine. In *Conservation geography: Case studies in GIS, computer mapping, and activism,* edited by C. Convis. Redlands, Calif.: ESRI Press.

Schick, R. S., M. E. Lutcavage, and J. Goldstein. *Spatial patterns in bluefin tuna distribution in the Gulf of Maine (1993–1996).* In preparation.

GIS

The Key to Research Integration at the Australian Institute of Marine Science

AUSTRALIAN INSTITUTE OF MARINE SCIENCE

Stuart James Kininmonth
The Australian Institute of Marine Science
Townsville, Queensland, Australia

The establishment of an enterprise GIS at the Australian Institute of Marine Science (AIMS) is revolutionizing the acquisition of knowledge in the field. For the last 30 years, AIMS *(www.aims.gov.au)* has focused on building a knowledge base of the complex marine ecosystems of the tropics with research effort concentrated in the northern waters of Australia. However, the nature of modern research requires considerable specialization and has the tendency to generate isolated projects. The natural systems being examined are complex, with abiotic and biotic components, and thus highly interactive. Research projects address this problem by dissemination of their results in the final stages, but often this is too late for experimental integration. The data formats describing the research findings are usually custom-made for the particular research field, and direct comparisons are difficult. This represents a considerable waste of opportunity, especially in a multidisciplinary research organization such as AIMS. Efforts to address the interaction between projects utilize the conference and seminar mediums with mixed success. A unifying system based on a fundamental research element is required. That element is location and the system is an enterprise GIS (or EGIS).

An EGIS utilizes a centralized data management system with all data associated with a spatial feature. The guiding principles of standardization, accessibility, ease of use, and integration are determining the construction of the AIMS EGIS. This involves eventually placing all research data into a central relational database (Oracle 8i) controlled by ESRI's ArcSDE™ (Spatial Data Engine™). Having common data with useful metadata is critical to the success of this system. The use of Web-based mapping using HTML and Java™ generated by ArcIMS® (Internet Map Server) provide rapid access to the research results in a familiar Internet environment. Traditional ArcInfo workstation and the newer ArcMap™ 8.1 are available for all staff through a server-based license manager.

Integration of a project's activities occurs at three levels. First, a diverse operating environment at AIMS demands a system that can provide seamless information management. AIMS is unique in the research field with marine operations based on two ships that conduct scientific activities for 74 percent of the year. Consequently the widespread use of nautical charts as a base layer

PHOTO COURTESY OF SALLY INGLETON

Katharina Fabricius checks her location prior to working on her experiment on coral settlement and reef recovery.

to display project areas of interest, experimental sites, and research findings (such as water temperature from remotely sensed data) on board the ships and, importantly, back in the office, provides efficient integration of common data sets. With 351,400 square kilometers of Great Barrier Reef in which to conduct scientific activities, the use of Global Positioning System (GPS) and electronic charts is essential, and replaces the desperate, repetitive searches of past years. On a recent field trip, coral settlement plates were placed close to another experiment to maximize the use of valuable ship and scuba time, despite being dispersed over 1,200 kilometers. Back at the laboratory, the distribution of experimental sites can be displayed with the same data available at sea. Time-consuming data transfer and translation are avoided.

The second level of integration involves researchers comparing and analyzing their own data with the results of others. AIMS has a diverse array of research interests, including oceanographic modeling, marine biodiversity characterization, marine resource management, human impact mitigation, marine biotechnology, and technical engineering development. Using ArcIMS, all research staff have the opportunity to view and integrate data sets through standard, Web-based mapping. For instance, a geneticist investigating the thermal tolerance of coral (photo below) can examine the sea temperature history collected by other project groups with a few mouse clicks. The long-term monitoring group members have been leaders in this area, with results immediately available on the Web after each voyage. Between dives the

Mary Wakefield stereo-photographs coral at Young Reef for a monitoring study conducted over 20 years.

PHOTO COURTESY OF TERRY DONE

PHOTO COURTESY OF STUART KININMONTH

Carolyn Smith and Ray Berkelmans compare the genetic and physiological basis for thermal tolerance of corals.

survey information is collated to a specified standard in a relational database on board the ships. The age-old problem of motivating staff to prepare and submit data *to a standard* is solved by the open visualization of their data. More sophisticated modeling by specialist staff requires the use of ArcInfo, Idrisi, and custom-made applications. In particular, the use of Bayesian statistical methods to interrogate the multiple data sets highlights the opportunities that integration presents.

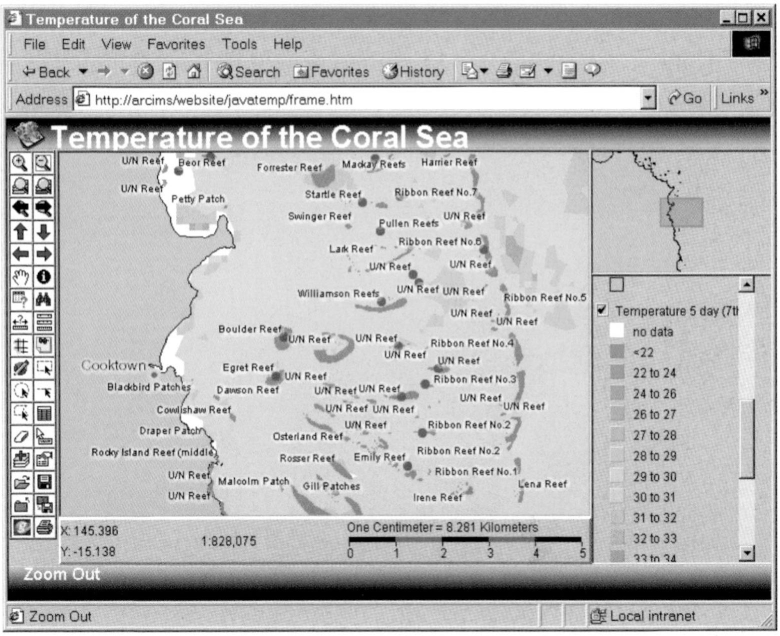

Figure 1: Screen capture of a Web map showing coral bleaching survey data with reefs and remotely sensed temperature data.

A fully functional GIS is a tremendous benefit to an institute, but many data sets are nonspatial yet are still important to the scientific endeavors at AIMS. In a strategic initiative to enhance the entire utility of information at AIMS, the development of a sophisticated data center incorporating the essential activities of operation support, corporate management, and nonspatial science data is in progress. This new data center will utilize the spatial services of the EGIS. The final result will be a system where perhaps the budget information could automatically be an attribute of a coral-spawning experiment.

The third level of integration is the complex sharing of research knowledge between organizations. Recognizing the benefit of synergetic organizational structures, the Australian federal government established cooperative research centers (CRC). The CRC Reef is a "knowledge-based partnership that provides research solutions to protect, conserve and restore the world's coral reefs..." (CRC Reef mission statement, *www.reef.crc.org.au*). Using the growing utility of the Web for data sharing, the CRC Reef is supporting the development of Internet mapping and hyperlink information networks. These easy-to-use mass media tools facilitate the exploration of complex issues through informative displays generated by the relevant organization. Maps that permit exploration and simple analysis are essential to this process, where the complexity of the issue is not masked by a textual summary *(figure 1)*.

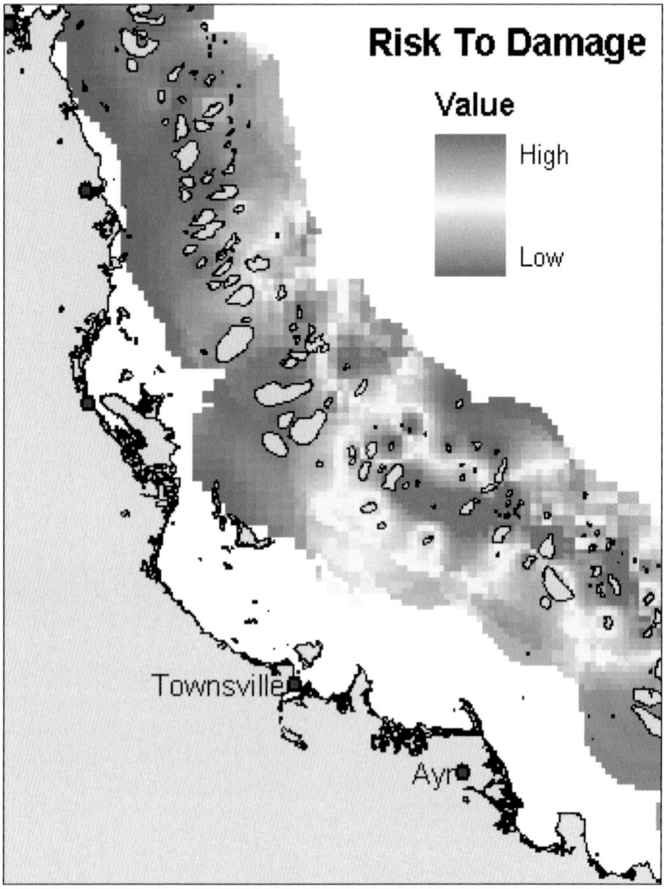

Figure 2: Map generated by the Bayesian "weight of evidence" model of the predicted damage of crown-of-thorns starfish by Scott Wooldridge.

With the recent installation of spatial Web servers in other marine organizations, the ability to combine up-to-date spatial information in a customized format is a powerful addition to the effectiveness of knowledge dissemination. The display of research results is rarely hassle-free, and the protection of intellectual property rights remains a difficult issue. The livelihood and future prospects for a researcher rely on the exclusive rights to scientific data until published. This process often takes several years, by which time the use of the raw data by another researcher may be temporally too disjointed. To address this issue in the short term, researchers can supply summary results that will facilitate integration. In the long term, a negotiated period is established after which the data becomes available for open institutional use. Experience with global coral species data shows that the popularity of a data set can be influenced by the open promotion of summary data such as a visual portrayal of extrapolated point samples (see also Shapiro and Varian 1998).

The hunger for visual displays highlighting the integration of past events is considerable, but the thirst for visual representations of the future is insatiable. The CRC Reef-funded group at AIMS, called "Reef Future," seeks to use modern statistical methods to explore future scenarios based on present actions. Utilizing a vast array of spatial data and model outputs, Reef Future researchers such as Terry Done, Glenn De'ath, Scott Wooldridge, and the author are value-adding 30 years of research results. Using Bayesian "weight of evidence" models combined with predictive statistical procedures, the future status of selected tropical ecosystems can be described *(figure 2).* This form of modeling can be achieved only with data sets that are professionally managed in an EGIS.

EGIS is still in its early stages of development and many valuable data sets remain to be converted, but the momentum to adopt modern spatial tools in support of the exploration of the marine environment continues to increase.

Acknowledgments

Special thanks to the information technology team (Scott Bainbridge, Peter Hewitt, James Smith, Carsten Malan, and Cathryn Anderton) for supplying the wonderful system that supports the institute's EGIS.

Reference

Shapiro, C., and H. R. Varian. 1998. *Information rules: A strategic guide to the network economy.* Cambridge, Mass.: Harvard Business School.

Moored Buoy Site Evaluations

Roger A. Goldsmith and A. J. Plueddemann
Woods Hole Oceanographic Institution
Woods Hole, Massachusetts

The Woods Hole Oceanographic Institution (WHOI) Upper Ocean Processes Group routinely deploys instrumented surface buoys that collect upper-ocean data in support of air–sea interface science programs. Mooring a buoy to the ocean floor, often in water several miles deep, keeps the instruments in the same location, enabling measurement of the environment at a remote location over extended periods of time. The surface buoy is equipped with meteorological sensors, and the subsurface mooring line, connecting the buoy to an anchor with an acoustic release, is often outfitted with additional oceanographic instruments. Researchers can thus associate events at the surface with changes in the water mass below.

A recent project called for an initial one-year surface-mooring deployment at the Northwest Tropical Atlantic Station (NTAS), on the southwest flank of Researcher Ridge (approximately 15° N, 51° W) in 5,000 meters of water. The area is located above the abyssal plain, northeast of South America, in a region where there is significant variability in the sea surface temperature anomaly. The surface buoy measures parameters associated with the ocean's response to events at the air–sea interface, and uses the information to assess models of heat and moisture fluxes.

Deep ocean mooring deployments provide several challenges, including planning the details of the mooring design, and the tolerance for error in the deployment depth. Moorings are typically of compound construction, consisting of chain, wire, and synthetic elements. Principal design considerations are the compliance (degree of "stretch"), resonance (the natural frequency of vibration), and scope (ratio of mooring length to water depth). These three elements are "tuned" to achieve the desired static and dynamic response characteristics. Careful tuning can provide a more robust mooring, but for the tuning to be effective the actual deployment depth must match the design depth to within a few percent (e.g., within 100 meters for a 5,000-meter design depth).

This implies that the bathymetry surrounding the chosen deployment site must be well known, and that the mooring placement must be relatively precise. The availability of accurate global bathymetry on scales of a few hundred kilometers provides a valuable starting point for determination of the mooring site. However, information on smaller scales is desired for planning the deployment, and a shipboard bathymetric survey is typically done to refine the site.

It is desirable to have a target area of 1 square kilometer or more due to difficulties in placing the mooring precisely. The mooring is deployed buoy-first, and strung out behind the ship while steaming forward slowly (for example, at 1 knot). When the anchor is released, it does not drop straight down. Instead, the drag of the mooring line and instrumentation serves to "pull" the anchor toward the buoy during its descent. Anchor "fall-back" is typically about 5 percent of the water depth, but varies with the mooring design and instrument load. Wind and currents influence the angle at which the mooring streams behind the ship, and thus the direction in which the anchor falls back. Finally, vagaries of the ship's speed and track through the water during the deployment process mean that the exact anchor release location is not predictable. Overall, these factors introduce uncertainty of at least several hundred meters in the actual anchor position relative to the target position.

To minimize the time necessary for the shipboard survey, and to maximize the likelihood of meeting the design depth tolerance, it is useful to focus on regions surrounding the target depth where the bottom slope has a local minimum. Clearly, this becomes more important when the local topography is steep (e.g., the flank of Researcher Ridge). To facilitate the mooring deployment described here, a GIS application was used to identify portions of the southwest flank of Researcher Ridge within plus or minus 100 meters of the 5,000-meter isobath and with local slopes less than 10 degrees.

This type of topographic site feasibility study is done all the time in civil engineering applications, but this small project provides a good example of how existing GIS software and techniques can be adapted for use with ocean-ographic research. As a starting point we used the 2-minute bathymetry derived from satellite gravity measurements (Smith and Sandwell 1997). As this was a deep, mid-ocean location, the approximately 3.7-kilometer horizontal resolution was adequate for our investigation. In some areas, higher-resolution bathymetry data may be available from hydrographic or multibeam surveys and could be used if available.

Because the general area was only about 2 degrees (200 kilometers) square and the bottom slope was a factor in the siting criteria, we converted the grid to 2-kilometer cells and used a Universal Transverse Mercator (UTM) projection. We used an ESRI ArcView software application specifically for computing surface slopes. Another common GIS procedure, the generation of buffers, was employed to define a zone within 5 kilometers of the 5,000-meter isobath. When these preliminary steps were completed, it was a straightforward procedure to identify all the areas meeting the defined criteria: within 5 kilometers of the 5,000-meter depth, between 4,900 and 5,100 meters depth, and less than 10-degree bottom slope. These areas are shown in red in the accompanying figure 1.

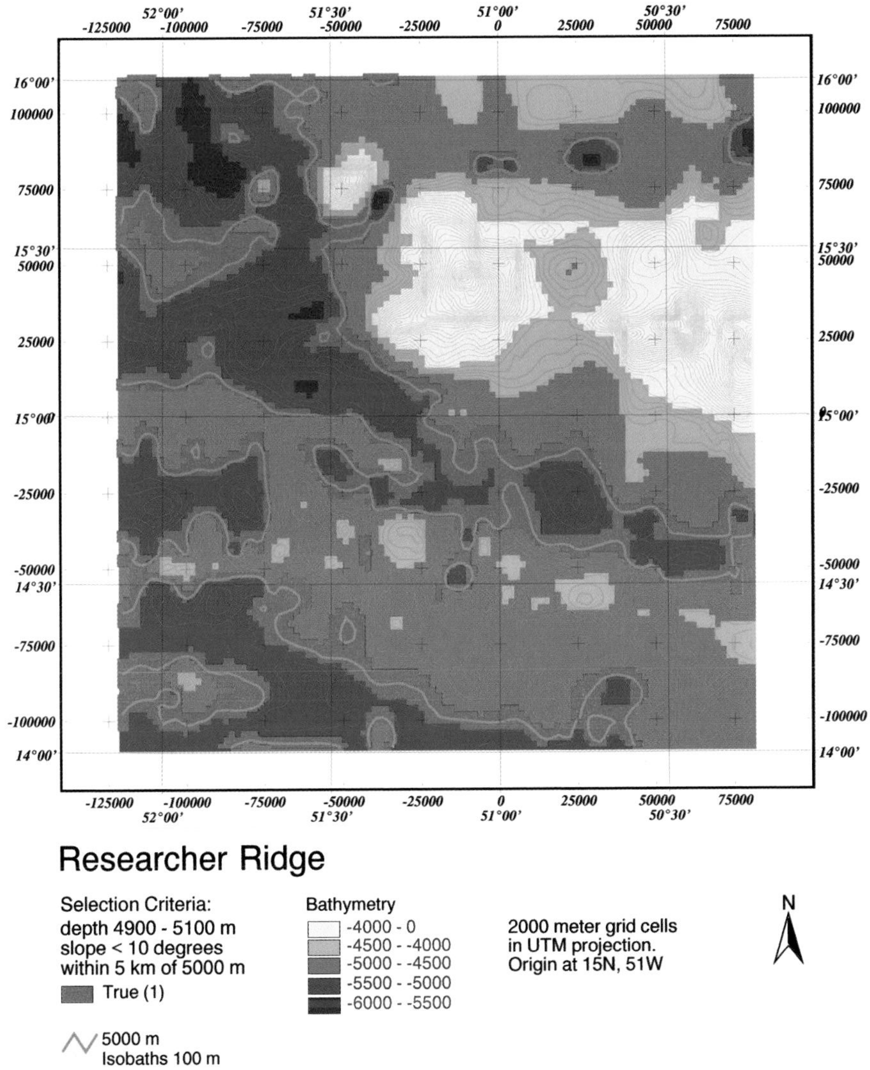

Figure 1: This raster representation of the ocean floor incorporates both the depth and slope data used to indicate where on Researcher Ridge to place the buoy.

A tentative site was identified very close to the chosen grid origin at 15° N, 51° W. In this case the site was in the vicinity (25 kilometers south) of what had been identified by simple visual inspection of a nautical chart. An enlarged view of the area is shown in figure 2. A proposed cruise track was drawn in to help conduct the detailed survey, and a chart was prepared showing both the Universal Transverse Mercator (UTM) grid and geographical coordinates useful for navigation. The entire exercise was done using existing GIS software, tools, and commonly available data. The procedure lends itself to being done at sea in an operational environment. Here it not only helped refine the siting of the mooring location but also identified several alternative sites that might not have been so readily apparent.

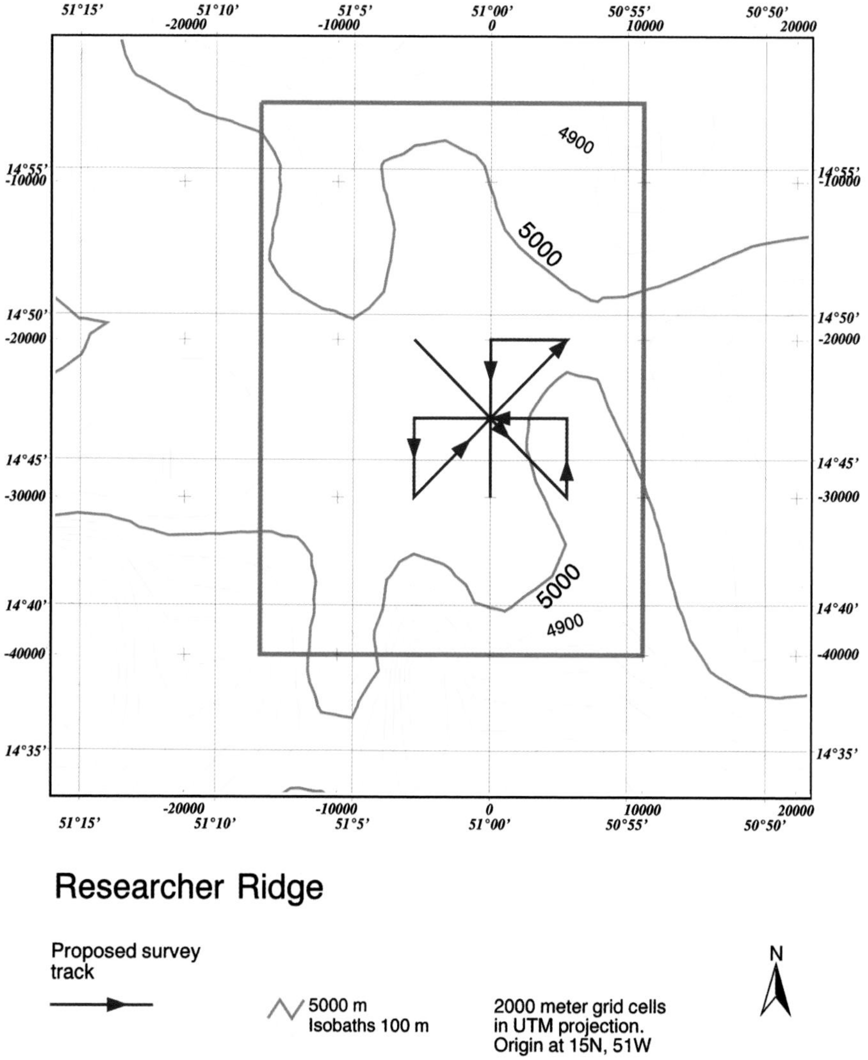

Figure 2: X marks the spot? Using ArcView, this chart was created to assist in planning the cruise track for deployment and mooring the buoy.

Postscript

On March 30, 2001, the crew of the R/V Oceanus, Cruise 365, Leg 5 conducted a bottom survey in an "X" pattern over the selected area. The region was found to be surprisingly flat, with a depth of about 4,980 meters at the center and a variability of only +/–60 meters. The mooring was successfully set at 14.83283° N, 51.005° W in 4,982 meters of water.

Reference

Smith, W. H. F., and D. T. Sandwell. 1997. Global seafloor topography from satellite altimetry and ship depth soundings. *Science* 277 (September): 1,957–62

11 Mapping Bottom Habitats Along the Southeastern U.S. Continental Margin

Douglas Wilder[1] and Henry Norris
Florida Marine Research Institute
Florida Fish and Wildlife Conservation Commission
St. Petersburg, Florida

Hard-ground and reef areas in the marine ecosystem provide essential habitat for a broad variety of marine species, including sea turtles, lobsters, and an abundant array of fish and invertebrate species. The effects of commercial and recreational fishing, as well as myriad other anthropogenic pressures, can threaten the long-term viability of fish populations. Study and management of these effects are problematic because of the difficulties involved in assessing fish population size and flux. Knowledge of the location and extent of critical habitat allows researchers and managers to track more accurately the effects of fishing and thus more effectively protect essential areas. This knowledge is crucial to the protection of reef-type habitats and the practice of sustainable fish harvesting.

The Southeast Area Monitoring and Assessment Program–South Atlantic (SEAMAP–SA) established the Bottom-Mapping Workgroup in 1985 to gather, archive, and disseminate hard-bottom habitat data needed by researchers and managers to study and protect essential fish habitat (EFH). The group was given the task of producing a regional database containing seafloor habitat information on the continental shelf areas from the North Carolina–Virginia border to Jupiter Inlet in Florida (just north of West Palm Beach). Principal components of the database include location, extent, and general type of bottom habitat. Because hard-bottom reef areas are the preferred habitat for a wide variety of commercially and recreationally harvested fish species, the SEAMAP–SA database was built around data representing these areas. The Bottom-Mapping Workgroup was also directed to establish protocols for collecting and archiving bottom-mapping data as well as for characterizing bottom type based on the collected data. To enable the database to become a powerful tool for assessing EFH, it needs to be flexible and facilitate queries and browsing.

The Bottom-Mapping Workgroup has established protocols for data collection and archiving.

The four major objectives identified to fulfill the workgroup's role are to:

- Conduct an extensive search of existing databases and identify data records that could be used to classify where hard-bottom reef habitats were present on the continental shelf of the South Atlantic Bight from Florida through North Carolina and from the beach out to a depth of 200 meters.
- Analyze each database using standardized protocols that define whether a data record provides clear evidence of reef habitat, possible evidence of reef habitat, or no evidence of reef habitat, and to identify the locations of artificial reefs.
- Summarize the bottom-type information into flexible, easy-to-use databases that will provide researchers and managers with pertinent information concerning the location and extent of these areas, types of data used in determining bottom type, and source of the data.
- Convert all data to a geographic information system (GIS)-readable format.

Van Dolah et al. (1994) completed the first of three bottom-mapping publications reporting on surveys covering the South Atlantic Bight (SAB). This publication reported on data for shelf areas off South Carolina and Georgia and was preceded by establishment of data-formatting and -evaluation protocols through a series of intensive workshops conducted by the Bottom-Mapping Workgroup. Moser et al. (1995) cataloged and described available bottom-type data for shelf areas off North Carolina, and Perkins et al. (1997) cataloged bottom-habitat data in the shelf areas off Florida southward to Jupiter Inlet. Initial Bottom-Mapping Workgroup reports included evaluations of GIS packages for browsing, querying, and displaying bottom-type data. In 1997, the committee decided to extend the study area southward through the Florida Keys. The additional mapping and cataloging of the Florida Keys data were completed by the Florida Fish and Wildlife Conservation Commission's Florida Marine Research Institute (FMRI) and published on CD–ROM in 1998. The CD–ROM included previous workgroup reports in digital format and provided, in a GIS-browsable format, a seamless version of the workgroup's database describing the bottom habitats from North Carolina through Florida. A second CD–ROM was produced in 1999, and a third was incorporated into a compiled paper-copy document of the three original reports in 2001. Figure 1 depicts the survey areas covered by each report.

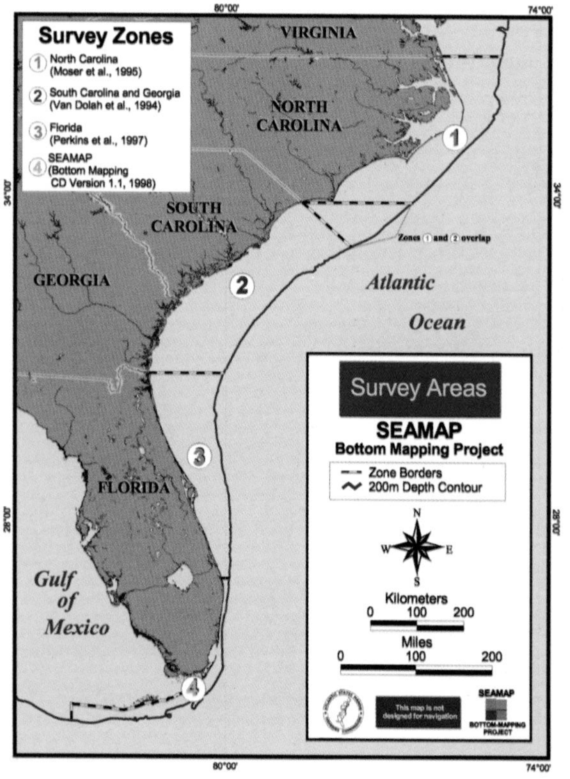

Figure 1: South Atlantic Bight areas surveyed during the SEAMAP–SA Bottom-Mapping Project. Zones 1, 2, and 3 were the subjects of three documents. Zone 4 data was included on the SEAMAP–SA bottom-mapping CD–ROM.

Methods

The Bottom-Mapping Workgroup developed the format for the SAB database prior to data collection. During a series of workshops attended by representatives from the four coastal states, the National Marine Fisheries Service, and South Atlantic Fisheries Management Council, a common format was developed. Historical sources were then identified, analyzed, and considered for inclusion in the SAB database. Much of the identified data (from more than one hundred studies, some dating from the 1930s) was not intended for use in a GIS. Positional information had been collected via various methods, including dead reckoning and Loran C. A protocol for determining positional accuracy and protocols for deducing bottom type, based on the type of gear and the method used in collecting each sample, were established. Gear types and methods used in these historical surveys included sidescan sonar, vibra core, aerial photography, dredge, trawl, and trap. The workgroup also established protocols for tabulating the various data sources and relating them to the primary bottom records.

Categorizing the data

Given the wide array of bottom-type descriptions used in various studies of the continental shelf, it was necessary for the workgroup to devise common classifications for the different bottom types. Using criteria based on the type of equipment used in a given original study, the workgroup classified bottom types as Hard Bottom, Possible Hard Bottom, or No Evidence of Hard Bottom. The Hard Bottom category was further divided into Artificial Reef and Hard Bottom on Artificial Reef. Hard bottom data includes any natural reef habitat as well as nonliving hard ground areas. Possible hard bottom areas are those where a definitive determination of hard bottom is not possible but evidence suggests its presence.

For much of the data, a determination of bottom type depended on the species of fish captured in trawls or traps. Van Dolah et al. (1994) published a list of reef-obligate species used to determine bottom types in South Carolina and Georgia shelf waters. Moser et al. (1995) expanded the list to include fish species in North Carolina waters, and Perkins et al. (1997) added fish species found in Florida to the list. Prior to publication, each list was subjected to the review and approval of the workgroup. This process resulted in an effective protocol for determining the locations of hard-bottom habitat: examination of trawl and trap data for the presence of those fish species considered to be indicators of the presence of reef-type habitat. The list includes 245 species, 17 genera, and four families. Careful attention had to be given to those fish species (e.g., various apogonids, lutjanids, and haemulids) whose habitat use depends on location (Hardy 1978; Gilmore et al. 1981 in Perkins et al. 1997). These species may be found in grass beds in shallow, inshore waters, but are found only in hard-ground areas in deeper waters (Perkins et al. 1997). Consideration was also given to species deemed hard-bottom obligate in shelf waters off one state but not off others. Such latitudinal habitat-use variation is observed in several species that occur within the SAB.

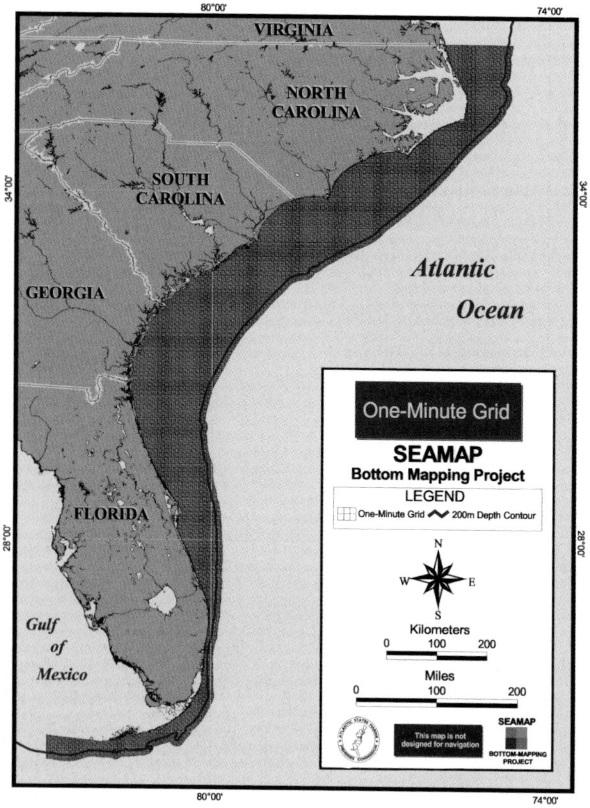

Figure 2: The extent of the 1-minute grid (cell size = 1 minute × 1 minute) is used to compile SEAMAP–SA bottom data. Of the 51,956 cells, 14,674 contain information on bottom type.

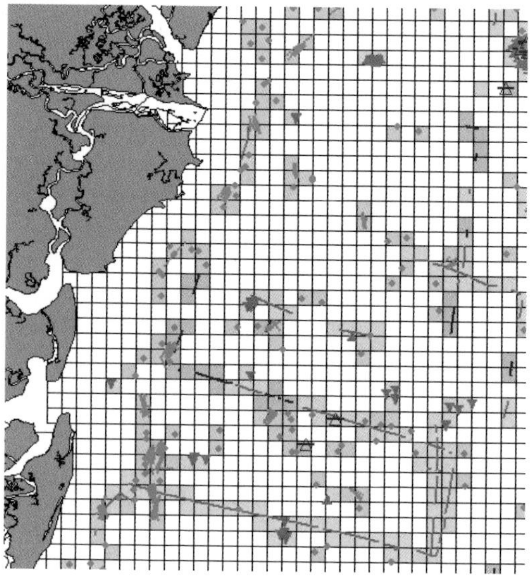

Figure 3: Each grid cell is Hard Bottom, Possible Hard Bottom, or No Evidence of Hard Bottom. Data locations falling on cell boundaries are assigned to the north- or west-adjoining cell. Data that crosses more than one cell (transect data) is generally assigned only to the cell in which the transect begins. However, certain transect data, such as sidescan and others, is assigned to each cell crossed.

To further complicate the issue, several species are known to use different habitats at different life stages. For example, many species in the families *Lutjanidae* and *Balistidae* use grass beds or pelagic habitats as nursery areas (Hardy 1978; Bortone et al. 1977 in Perkins et al. 1997) but are considered reef species as adults (Perkins et al. 1997). Because much of the collected bottom data did not contain information on the size (life stage) of the specimens caught, this particular situation is not addressed. Therefore, the reef-obligate species list does not include species that use hard-bottom habitats only during certain life stages. Locations where reef-obligate species were caught were classed as Hard Bottom.

Organizing the data spatially

According to the protocol established by the Bottom-Mapping Workgroup, the SAB is divided into 1-minute latitude by 1-minute longitude grid cells used to organize the data *(figure 2)*. Each data record is assigned a unique number as well as a "block" number indicating the latitude and longitude of the southeast corner of the grid cell in which it falls. Record locations falling on the boundary between grid cells are assigned to the cell to the north or the west of the record location *(figure 3)*. A routine was written in ArcView to assign the block numbers to each data point based on each record's coordinates.

In the case of line or polygon data, the SEAMAP–SA data protocol calls for a different approach in assigning block numbers. The protocol stipulates that each transect (e.g., data collected by trawl survey, some dredge surveys, certain sidescan sonar data, and some diver observations) assign a bottom type to the grid cell in which the transect begins. However, certain transect data allows for more accurate designations of bottom type (e.g., designations made by using sonar tracks and diver observations) and therefore code each cell through which these transects pass. Polygon data (e.g., data collected via sidescan sonar and aerial photo interpretation) is assigned to each grid cell crossed. A given reef or hard-bottom assemblage may intersect several grid cells; whenever this occurs the database records a unique record for each cell in which the assemblage falls (each cell that a reef area overlays). Regardless of how transects and polygon data are divided into the grid, each transect or area maintained its identification as a single entity in the database.

The result is an expansive grid (51,956 cells, 14,674 with data) of 1-minute cells covering the SAB, with each cell recording the number of data records and bottom type that occurred within the cell and, through the database, tying back to the original data source. Source information includes specific study or project names, agency, or organization contacts, and dates of collection. Each grid cell was assigned an overall bottom type based on a hierarchical scheme developed by the workgroup. A cell with any Hard Bottom data is automatically assigned a bottom type of Hard Bottom regardless of other data found within the cell. Similarly, if a grid cell contains Possible Hard Bottom data, and no instances of Hard Bottom, it is coded as Possible Hard Bottom regardless of any No Evidence of Hard Bottom data also in the cell. A cell with only No Evidence of Hard Bottom data is coded as No Evidence of Hard Bottom, and cells with no data are not classified. The reason for this hierarchical classification is to ensure that all instances of hard bottom are identified.

A grid cell with any Hard Bottom–type data is assigned as such regardless of other data recorded in the cell.

Database

One of the significant achievements of the SEAMAP–SA Bottom-Mapping Workgroup was the successful amalgamation of data from numerous studies into a single comprehensive database. The database seamlessly archived a total of 125 disparate data sets from a wide variety of sources, including government agencies, private contractors, and the research community. The database began with work completed by Van Dolah et al. (1994) for shelf waters off South Carolina and Georgia. Subsequent SEAMAP–SA reports (Moser et al. 1995; Perkins et al. 1997) contributed data for North Carolina and Florida. Additional data from South Carolina waters and the Florida Keys was added when the first Bottom-Mapping CD–ROM was published in 1998.

Table 1: Summary of bottom-mapping records by state and bottom type.

State	N. Carolina	S. Carolina	Georgia	Florida	Totals
Bottom type: Hard Bottom	2,006	4,414	1,206	14,058	21,684
Possible Hard Bottom	1,527	1,261	894	3,292	6,974
No Evidence of Hard Bottom	9,244	5,700	1,664	19,648	36,256
Artificial Reef	113	147	119	312	691
Hard Bottom on Artificial Reef	0	12	0	2	14
Not applicable	0	0	3	105	108
Total number of records	12,890	11,534	3,886	37,417	65,727

To date, 65,727 records have been compiled in the database. Table 1 summarizes the bottom-type records collected in each participating state. Source, date, location, gear type, and several other parameters are included for each data record via tables related by "block" number or an agency code unique to each data source. To facilitate use of the data, the Bottom-Mapping CD–ROM integrates the database with ArcView shapefiles and a customized ArcView project file.

Mapping and CD–ROM development

A series of 30 maps, each covering roughly 1 degree of latitude and longitude, depicts the bottom-mapping data in a static format. The data is also available in a browsable format on the Bottom-Mapping CD–ROM. Workgroup members at the Florida Marine Research Institute have developed three versions of the Bottom-Mapping CD–ROM. The latest is included with a new paper-based compilation of the three previously published Bottom-Mapping reports (Van Dolah et al. 1994; Moser et al. 1995; Perkins et al. 1997). The newly compiled document includes complete copies of the original reports, updated maps that reflect data added since these original reports were published (for South Carolina and Florida), and a new map series depicting the 1-minute grid coded to the bottom type determined by the data *(figure 4).* The accompanying Bottom-Mapping CD–ROM contains a customized ArcView project file for viewing shapefile versions of the database. The CD–ROM also contains digital copies of the publications that have been interlinked to allow easy navigation and searching. The CD is designed to provide scientists and managers with a tool for examining the bottom-mapping database and reports. The CD application may be used directly from the disk or installed to a local hard drive for greater integration with other data.

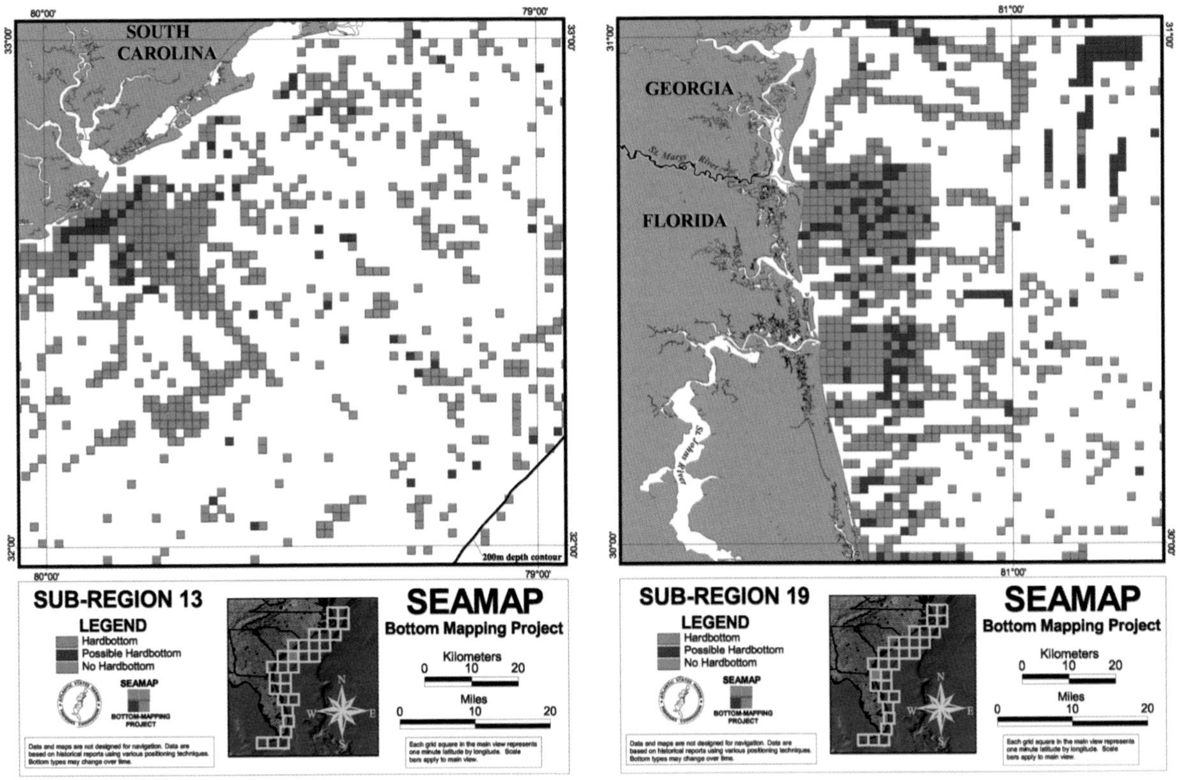

Figure 4: Examples from the map series depicting the 1-minute grid coded to a bottom type.

Data summary

Approximately 34 percent of the records in the SEAMAP–SA database are classed as Hard Bottom (including Artificial Reef and Hard Bottom on Artificial Reef), 11 percent are classed as Possible Hard Bottom, and 55 percent are classed as No Evidence of Hard Bottom. A total of 14,674 grid cells (28 percent of all grid cells) contain information on bottom type. The number of data records alone does not reflect actual distributions because many of the studies that generated the data focused on reef and hard-ground areas (Van Dolah et al. 1994). Included in the number of data records are data regarding reef areas in the Florida Keys (polygons) that have been divided into the 1-minute grid according to workgroup protocols. The number of records in the database reflects the number of cells crossed by such area data. The distribution of coded grid cells, however, offers a coarse view of natural hard-bottom distribution but does not reflect habitat-type proportions in the database because priority in grid cell coding is skewed to hard bottom. Although the bottom-habitat distributions recorded in the SEAMAP–SA database have not been rigorously examined, casual observations suggest that a majority of the reef-type bottom habitats occur (as expected) in the southern extremes of the study area.

Database use

The database and coded grid provide scientists and managers with a regional view of bottom types throughout the SAB. Detailed source information for each data record allows curious users to obtain more specific information about areas of interest. The database and grid also demonstrate historical survey activity in shelf waters, highlighting regions of intense study and areas for which little data has been collected.

The Bottom-Mapping Workgroup continues to search for data that could be added to the SAB database. The workgroup is also exploring expansion of database coverage to include deeper waters, which represent important habitat for several deepwater species including wreckfish, snowy grouper, yellowfin bass, slimeheads, and tilefish. The workgroup recently developed preliminary protocols to address the various characteristics of the ocean floor considered important to several of these and other species. Additional data assimilation and mapping will be completed as budget and time constraints allow.

Footnotes

[1] Doug Wilder is now with the National Park Service, Alaska.

References

Van Dolah, R. F., P. P. Maier, G. R. Sedberry, C. A. Barans, F. M. Idris, and V. J. Henry. 1994. *Distribution of bottom habitats on the continental shelf of South Carolina and Georgia.* Final report submitted to the Southeast Area Monitoring and Assessment Program, South Atlantic Committee.

Moser, M. L., S. W. Ross, S. W. Snyder, and R. C. Dentzman. 1995. *Distribution of bottom habitats on the continental shelf off North Carolina.* Final report to Marine Resources Research Institute, South Carolina Department of Natural Resources for the Southeast Area Monitoring and Assessment Program, Bottom-Mapping Workgroup.

Perkins, T., H. Norris, D. Wilder, S. Kaiser, D. Camp, R. Matheson, Jr., F. Sargent, M. Colby, W. Lyons, R. Gilmore, J. Reed, G. Zarillo, K. Connell, and M. Fillingfin. 1997. *Distribution of hard-bottom habitats on the continental shelf off the northern and central east coast of Florida.* Final report submitted to the Southeast Area Monitoring and Assessment Program Bottom-Mapping Workgroup and the National Marine Fisheries Service.

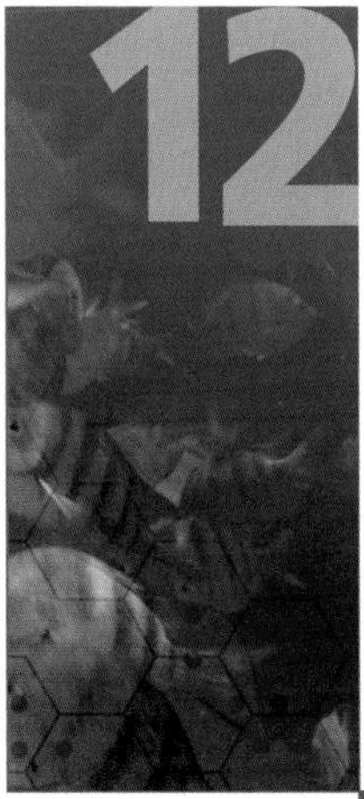

TerraMarIS—Terrestrial and Marine Information System

Matthias Mueller
University of Applied Science
GEOMAR Research Center for Marine Geosciences
Rostock, Germany

Bernd Meissner
University of Applied Science
Berlin, Germany

Wilhelm Weinrebe
GEOMAR Research Center for Marine Geosciences
Kiel, Germany

Many new scientific studies in marine-related subdisciplines have led to the availability of more detailed information in these fields. In addition, complex models within these fields often require the collection of huge quantities of data that are difficult to manage. Alternatively, the data exchanged among the disciplines may have to be limited to well-established research results, with large portions of the basic data remaining unused, even in multidisciplinary research settings. For example, scientists in the geology/tectonics/economic geology sector at the Research Center for Marine Geosciences (GEOMAR) in Kiel, Germany, have established a working group to promote interdisciplinary research of the seafloor and investigate the coastal zones and the interior of the adjacent mainland. Only a few studies prior to this have pursued similar goals (Bartlett 2000; Wright 2000). Researchers agree that the different disciplines focusing on terrestrial and marine investigations, respectively, need a tool at their disposal to allow their research results to be linked, analyzed, and visualized in the form of charts. It would also be useful to exchange not only final results, but large portions of the data sets acquired by the individual disciplines. It is expected that an interdisciplinary evaluation of data sets will allow a broader interpretation of the research results.

In cooperation with GEOMAR and the University of Applied Science Berlin, the rough concept of a terrestrial/marine geoscientific information system (GeoMarIS) was developed within the framework of this pilot study, to be applied to a small area on the west coast of Costa Rica. The project investigated the potential for integrating data from a large variety of sources. This approach was the basis for further development of this GIS tool within a larger framework.

The primary goal of this study is to transfer and archive the different data into a single, largely homogeneous data set, to achieve a suitable visualization of the data. However, the data is subject to the nomenclature and specific evaluation method of its particular discipline. The acqusition method and systems used vary for different data sets, as do the types of results or findings obtained. But the spatial reference is the common characteristic as a basic condition for the development of a GIS. The secondary goal is to create a tool for users allowing them to analyze, select, or simply obtain information via graphic displays of varying data.

Study area and databases

Figure 1: Cabo Blanco, at the southern tip of Nicoya Peninsula, with its adjacent continental margin.

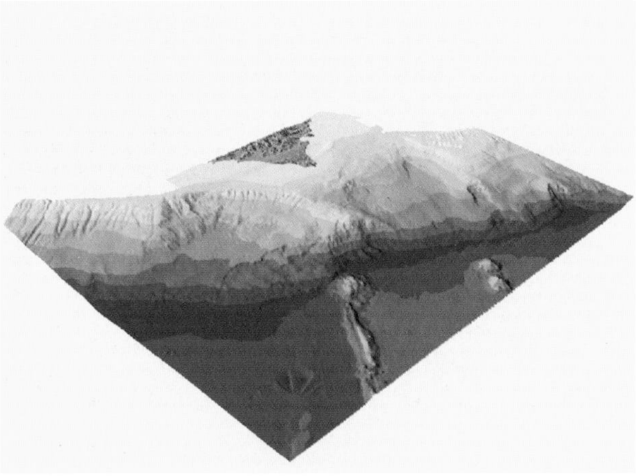

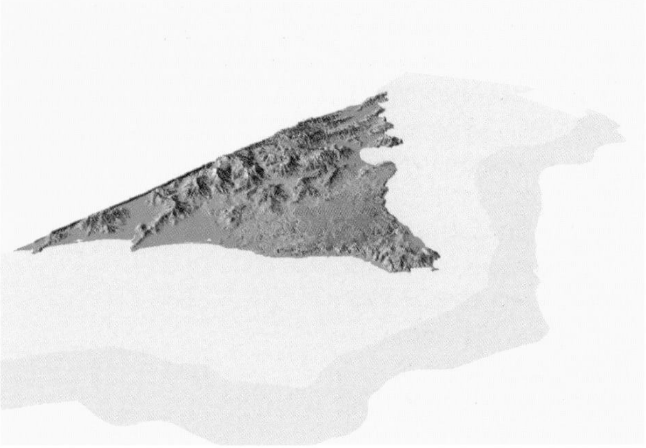

Figure 2: Digitized image of southern Nicoya Peninsula, with shallow water and terrestrial areas indicated.

The boundary of the study area, located on the Pacific Coast of Costa Rica, has been defined on the basis of GEOMAR's interests and activities *(figure 1)*. The area includes the Nicoya Peninsula with the Gulf of Nicoya and about 100 kilometers of open Pacific with the adjacent continental slope. The Nicoya Peninsula lies in the northeastern part of Costa Rica and has an area of some 5,000 square kilometers. It is part of the western margin of the Caribbean Plate and adjacent to the boundary between two lithospheric plates. The area is characterized by the subduction of the oceanic Cocos Plate, which moves northeastward below the adjacent Caribbean Plate with an average speed of 9 to 11 centimeters per year. The southern tip of Nicoya Peninsula represents a very interesting geological formation *(figure 2)*. The coastlines at this promontory of Cabo Blanco almost form a right angle (orthogonal) and mark the boundary between two different segments of subduction (von Huene et al. 2000). The oceanic plate south of Nicoya Peninsula is characterized by numerous seamounts, plateaus, and other topographic features, which, while subducted with the oceanic plate, create distinct trails on the overriding continental margin, such as Nicoya Slide, which was formed by several submarine slide events, and the Jaco Scar, a scar-shaped depression *(figure 3)*. Toward the coast, the subduction of seamounts cause elevation and rotation processes in the southern part of the peninsula (Fisher et al. 1998; Gardner et al. 2001).

In close cooperation with GEOMAR, data for the project was collected during several research campaigns on land and at sea under the guidance of GEOMAR. A geometric basis and descriptive data constitute the cornerstones of the terrestrial/ marine geo-information system TerraMarIS. The main activities in creating the geometric basis were the digitization of topographic maps (1:50,000) of the southern part of Nicoya around Cabo Blanco, which were subsequently converted to a digital terrain model (DTM). In addition, swath bathymetric data was linked to the DTMs as well as to Landsat ETM+ satellite imagery. Different databases were converted to uniform formats. A very important step was the transformation of data sets to common reference coordinates. In addition, extensive sets of GEOMAR volcanic and petrologic dredge data was used (Spangenberg et al. 1999). The goal of this dredging was to collect rock samples from the seafloor for subsequent petrological and geochemical analyses. A few to several hundred kilograms of sample material may be collected in this way.

Otherwise, rock samples from the Nicoya Peninsula represent the terrestrial part of the project data, within the framework of a larger project in western Costa Rica. The primary goals of the sampling were the age determination of igneous rock and geochemical analyses (Hauff et al. 2000).

Images from the Towed Ocean Bottom Instrument (TOBI) were incorporated into the GIS. TOBI is a deep-towed sidescan sonar system operated by the Southampton Oceanography Centre in the United Kingdom. It measures water depth, similar to a multibeam bathymetric mapping system, but also the hydroacoustic backscattering properties of the seafloor, thus revealing bottom characteristics for classification.

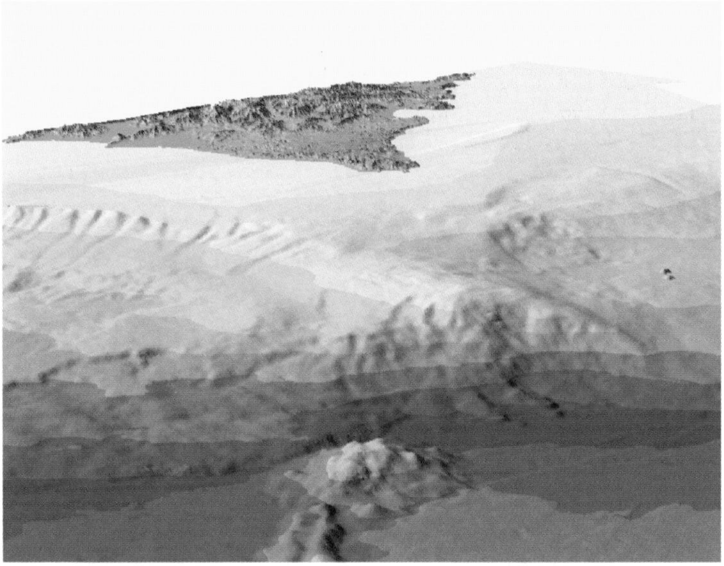

Figure 3: Cabo Blanco and adjacent bathymetry with characteristic topographic feature; view from southwest.

Data transformation and processing

Modeling of the geodata was carried out by means of ArcGIS 8.1. ERDAS IMAGINE 8.4 was used to process the satellite imagery and sidescan sonar data. The TerraMarIS project, including all geometric data, images, video sequences, and descriptive data, was implemented in ArcView 3.2. Among the most important activities in the realization of TerraMarIS was the transformation of a large variety of different data into compatible data, both with respect to the geometric basis and the descriptive information. After digitization and processing to obtain a DTM, this data had to be georeferenced to a common reference system with the echosounder data set and the Landsat scene. In this case, the UTM reference system was used, which is based on the reference ellipsoid of the World Geodetic System of 1984 (WGS 84). As the study area is located on the sixteenth meridian strip, a rectangular grid with metric units can be used for easier handling.

Unexpectedly, referencing to common coordinates posed major problems. The digitized topographic maps were available as geodetic coordinates in a Lambert conic projection with two standard parallels, using the 1866 Clarke ellipsoid as reference. The relatively small area of Costa Rica is divided into two parallel zones of latitude (north and south) with changing meridians. As the determination of a conversion factor for UTM coordinates was not possible initially, the Costa Rica values were converted to geographic coordinates. After subsequent transformation to UTM coordinates, a common reference system was arrived at for the TerraMarIS.

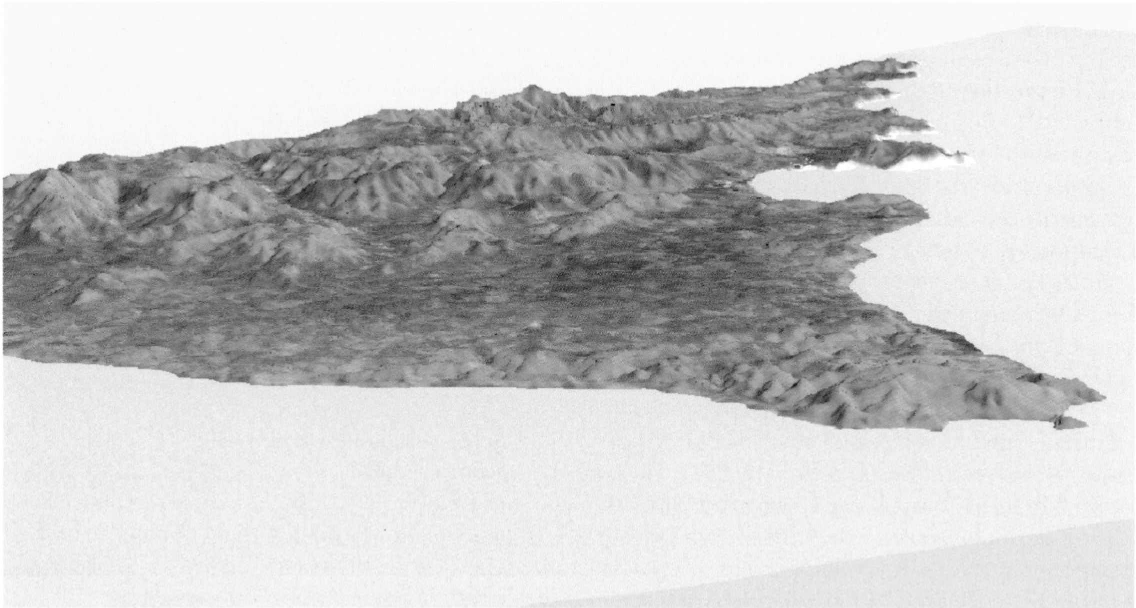

Figure 4: Cabo Blanco in a three-dimensional view with overlaid Landsat ETM+ subscene.

The multibeam echosounding data represents the depth of the ocean floor similar to topographic heights onshore. From the depth data a DTM is calculated, which is of special importance to marine geosciences, especially in areas of active geodynamic processes. The seafloor morphology is shaped by geological and tectonic forces. Determining the morphology helps to identify these processes and to reveal the geodynamic history. Bathymetric surveys thus provide more than just depth data (Weinrebe and Heeren 1997). The bathymetric database used here consists of several partial data sets from earlier research campaigns. The size of the x,y,z data set was found to be problematic with respect to processing and integration. By reducing the values and applying filtering functions it was possible to process the data and incorporate it into the project, together with the other geometric bases.

An important project component was the inclusion of satellite imagery. A Landsat ETM+ image from 2000 covers the extent of the area of the Nicoya Peninsula. The satellite imagery, in conjunction with the DTM data, served to ensure realistic visualization. Two subscenes were cut from the complete scene and incorporated directly into the project. One of them is used in a visualization of the complete Nicoya Peninsula, and the smaller section is used in a 3-D view, in combination with the Cabo Blanco elevation model that has been generated *(figure 4)*.

Besides geometric features, the descriptive data attributes constitute the basis of a GIS. It is the inclusion of the descriptive data and its spatial relationship with the objects that allows a GIS to be used for analyses, computations, and predictions. The data processed within the project was collected and evaluated in the course of geoscientific research projects by GEOMAR. Before the data could be entered into a relational database, it had to be transformed in such a way that its storage in a unified table structure was possible. The table structure is better accepted by the multidisciplinary partners involved, at least in the initial development phase of the database, because data sets can be easily identified in the table columns of the database. A transfer to an object-oriented database is planned at a later project development stage.

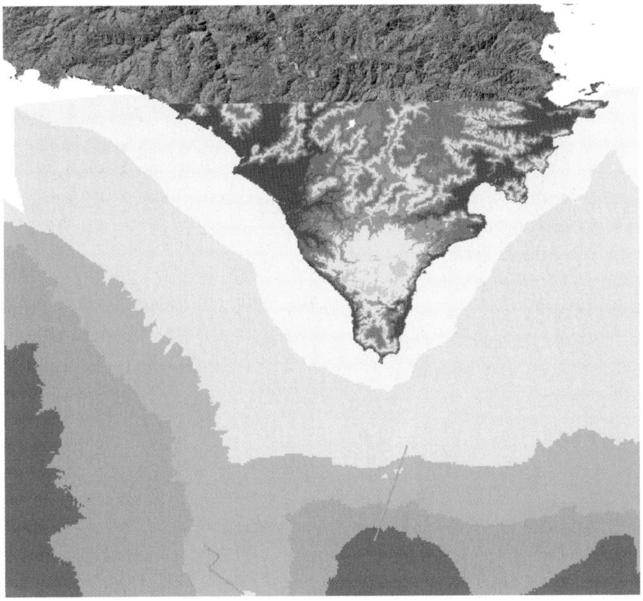

Figure 5: Landsat scene overlaid by the data in an ArcView shapefile.

Finally, metadata was also created and linked as a necessary step to trace back the origin and significance of heterogeneous data sets. Data was described using a metadata model with as much detail as possible. The concept of the metadata model is such that it can be expanded to allow later additions, and to meet changing user requirements. Despite the considerable effort involved in creating a metadatabase, it is indispensable if data sets collected by others are to be included in one's own investigations.

Preliminary results

The goal of the investigation is to integrate data from a highly heterogeneous background, both geometric and descriptive, into a largely uniform database. On that basis, a geoscientific information system was conceived and developed that will allow the analytical GIS tools to be applied to special scientific issues *(figure 5)*. It was found that integration of the data was a time-consuming process, especially the generation of a DTM from analog data and the processing of data into different data models. It has been found, however, that it is possible to combine geometric data from different sources. Nevertheless, it is impossible to completely eliminate differences resulting from different methods of data collection. This is apparent when visualizing the shallow water bathymetry around Cabo Blanco in comparison with multibeam echosounding data. The boundary between the two data sets can be clearly seen in the map layout of figure 6. The near coastal bathymetry there is characterized by less detail density than the more detailed deepwater part. White areas in the map example represent no-data fields. The boundary can be more or less dissolved by converting both data sets to raster format. Depending on the subject concerned, it is possible to use the advantages of either a vector or a raster data model.

The inclusion of data from geoscientific campaigns is a very important feature. Such data can now be easily displayed on one viewing plane and combined with geometric data as required *(figure 7)*. There is also the possibility of separating partial results from the tables via "queries," or to carry out new computations *(figure 8)*. Moreover, data sets can also be modified and enlarged.

The information content of GIS is additionally enhanced by including metadata, which allows for evaluation and comparison of the data quality. The metadata can be accessed directly through the menu. The user interface was designed using the ArcView scripting language Avenue, which allows users without previous knowledge of the available data sets to download the subjects directly into a view. Figure 6, once again, provides an example of a UTM grid with the complete extent of the Nicoya Peninsula as a Landsat image overlayed with topographic and bathymetric data. Red triangles show the locations of rock samples, orange dots indicate the dredge positions, and yellow lines represent the Ocean Floor Observation System (OFOS) traces.

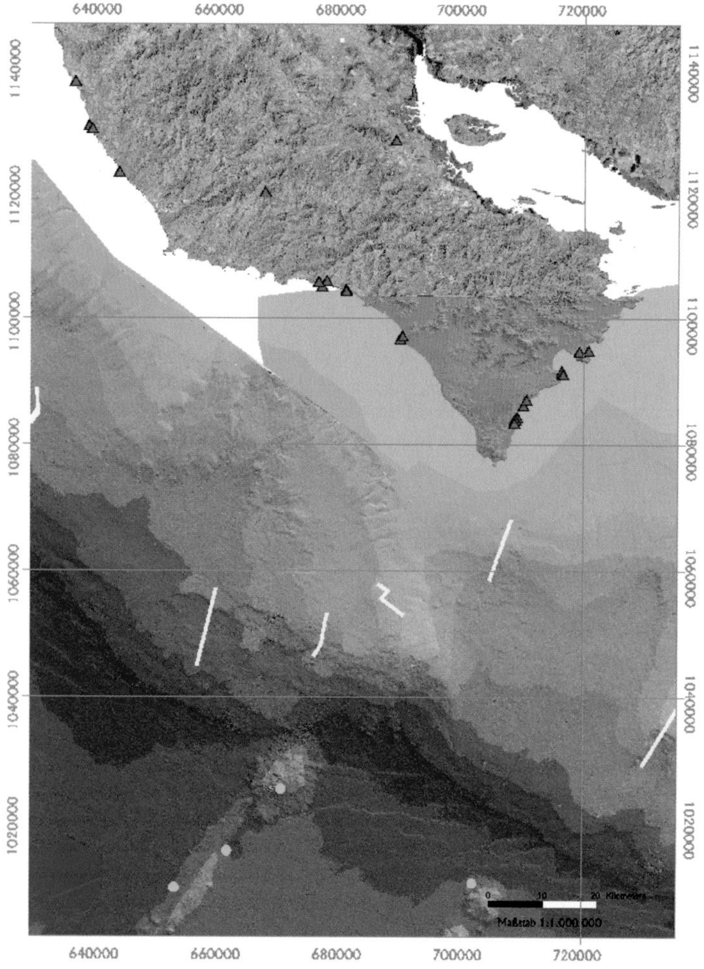

Figure 6: Example of a map layout showing the borderline between the multibeam data set and the digitized areas. Orange dots show locations of dredge activities, the yellow edges are OFOS tracks, and the red triangles are the locations of rock samples.

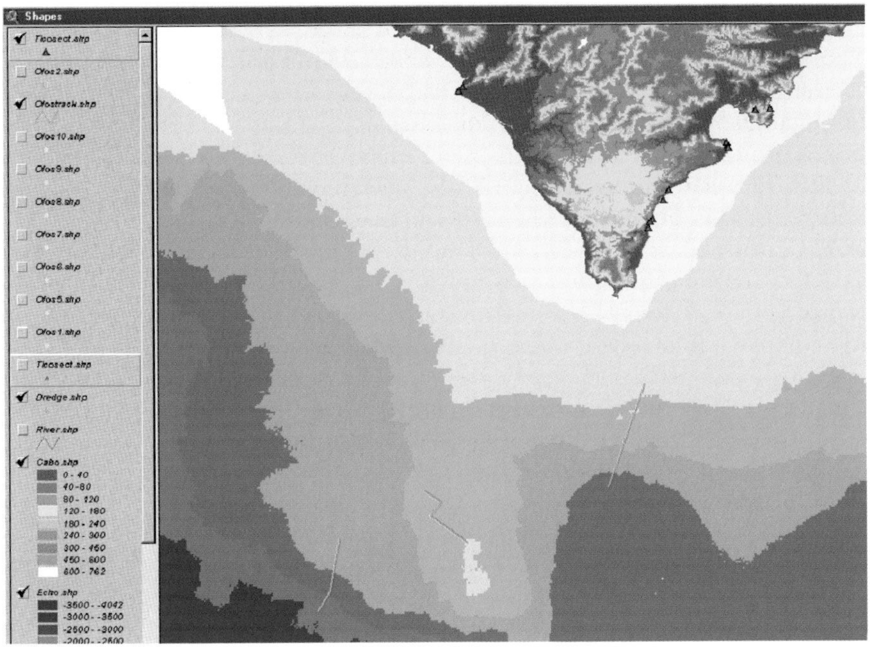

Figure 7: OFOS track lines are indicated by orange lines, and rock sample locations by red triangles.

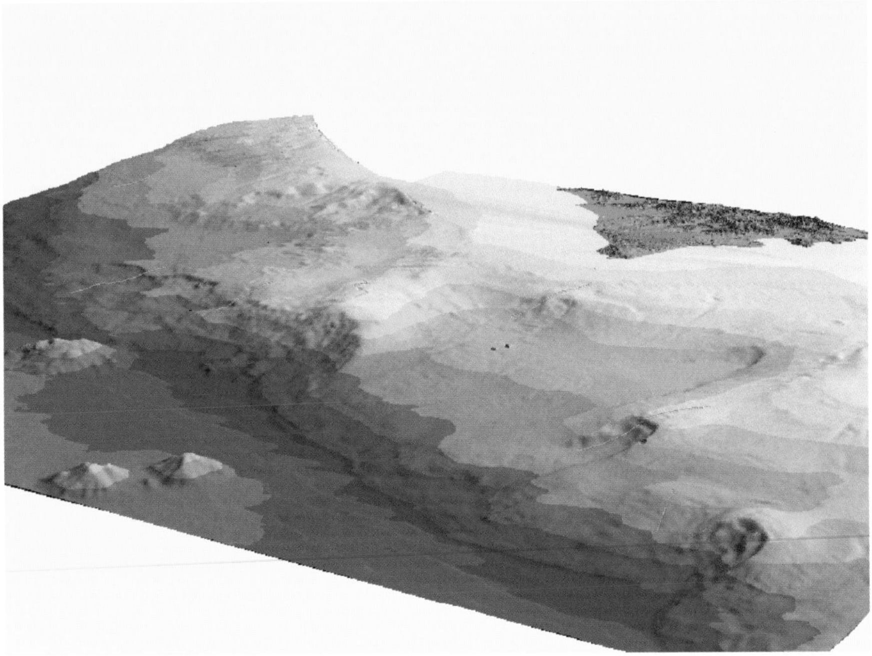

Figure 8: Overview of the digital data with the overlaid OFOS tracks.

Conclusion

The results of this pilot study prove that GIS-supported multidisciplinary research using available technology and methods is feasible. Existing data, information, and study results from different scientific disciplines have been linked, which has led to a definite improvement in the common representation of data from different sources, while also allowing an identification of existing knowledge gaps. For example, it has been found that sufficient bathymetric coverage of the shallow water area of Nicoya Bay is missing. To close this data gap, a portable multibeam system mounted to a tow frame developed at GEOMAR is planned. Operated from a small ship, this represents a cost-effective alternative compared to mapping with the large research vessels used so far.

By applying satellite imagery with limited field checks, it will be possible to complete the fracture tectonic pattern of the Nicoya Peninsula. Linkage of the terrestrial structures with those of the shallow and deepwater areas should allow an optimization of field work and a better understanding of the recent kinematics of the continental plate margins.

An improvement of the technical handling of data loading and analysis will certainly be necessary. Plans call for the continuation of this project to include not only tables and diagrams but also complete data sets, such as seismological surveys covering the entire area. An extension of the data sets to neighboring disciplines (for example, biological data with a geographic reference) would be another possibility. And an Internet connection via ArcExplorer™ and ArcIMS could further expand the scope of TerraMarIS.

Undoubtedly, GIS reaches its full potential only with very comprehensive data archives. It remains to be seen how the TerraMarIS project will perform in future practical applications. The quality of the information system could be improved by optimizing its individual components, such as database connections and user friendliness (e.g., Su 2000). Closer communication among the different disciplines will be desirable, especially with regard to the databases. This will also have a positive effect on the motivation of GIS developers. Although the effectiveness of a GIS depends on the creativity and expertise of its developers, the creation of a multidisciplinary tool requires much teamwork (Liebig 2001).

References

Bartlett, D. J. 2000. Working on the frontiers of science: Applying GIS to the coastal zone. In *Marine and coastal geographical information systems,* edited by D. J. Wright and D. J. Bartlett. London: Taylor & Francis.

Fisher, D. M., T. W. Gardner, J. S. Marshall, P. B. Sak, and M. Protti. 1998. Effect of subducting sea-floor roughness on fore-arc kinematics, Pacific coast, Costa Rica. *Geology* 26 (5):467–70.

Gardner, T., J. Marshall, D. Merritts, B. Bee, R. Burgette, E. Burton, J. Cooke, N. Kehrwald, and M. Protti. 2001. Holocene fore-arc block rotation in response to seamount subduction, southeastern Peninsula de Nicoya, Costa Rica. *Geology* 29 (2):151–54.

Liebig, W. 2001. *Desktop GIS mit ArcView GIS: Leitfaden für Anwender.* 3. Aufl: Wichmann Heidelberg.

Hauff, F., K. Hoernle, P. van den Bogaard, G. Alvarado, and D. Garbe-Schönberg. 2000. Age and geochemistry of basaltic complexes in western Costa Rica: Contributions to the geotectonic evolution of Central America. In *Geochemistry, Geophysics, Geosystems,* vol. 1, published by AGU and the Geochemical Society.

Huene von, R., C. R. Ranero, W. Weinrebe, and K. Hinz, K. 2000. Quaternary convergent margin tectonics of Costa Rica, segmentation of the Cocos Plate, and Central American volcanism. *Tectonics,* 19(2):314–34.

Spangenberg, Th., ed. 1999. Finalreport: Geologische und geophysikalische untersuchungen in seegebieten vor Costa Rica und Nicaragua—Beiträge zum verständnis des aktiven ostpazifischen kontinentalrandes. (Forschungsfahrt SO 107) Teil I, Textband.

Su, Y. 2000. A user-friendly marine GIS for multi-dimensional visualisation. In *Marine and coastal geographical information systems,* edited by D. J. Wright and D. J. Bartlett. London: Taylor & Francis.

Weinrebe, W., and F. Heeren. 1997. Hochauflösende bathymetrie: Basis mariner geowissenschaftlicher arbeiten. *Geowissenschaften* 15 (9):278–81.

Wright, D. J. 2000. Down to the sea in ships: The emergence of marine GIS. In *Marine and coastal geographical information systems,* edited by D. J. Wright and D. J. Bartlett. London: Taylor & Francis.

13 Multiple Marine Uses of GIS

Martin Kaye
Bay of Fundy Marine Resource Centre
Cornwallis Park, Nova Scotia, Canada

The use of GIS has long been the bastion of government as well as business. Today, marine geographical depiction through GIS is becoming a key factor in marine planning, policy making, conservation, and research. To begin the development process of integrated coastal resource management, the Coastal Resources Mapping Project for Digby and Annapolis Counties began as a joint project between the Western Valley Development Authority and the Department of Fisheries and Oceans (DFO)–Canada.

The affordability of today's GIS applications has allowed the Marine Resource Centre (MRC) to prioritize the development of a fully functional GIS lab. The MRC's Coastal Resource Mapping Project locally offers GIS information and technology that enable a wider use by community groups and organizations. Some examples of how the community uses MRC's data and maps include a Storm Surge Water Project in partnership with the Clean Annapolis River Project, Clam Flat Restoration/Harvesting Plan in partnership with the Annapolis and Digby Counties Clam Management Board, Coast Guard Risk Analysis Project in partnership with the Canadian Coast Guard (CCG) and, finally, the combined DFO and CCG Emergency Response Plan. The MRC's GIS project supports these causes in a visually meaningful way. The MRC continues to expand its in-house GIS information and to enlarge its accessibility to other groups involved in the region's integrated coastal resource management initiatives.

Because of the MRC's broad interest in integrated management of the Bay of Fundy and the Gulf of Maine, there was a need to expand some of the data sets for the entire Bay of Fundy regions and Nova Scotia coastal areas. Fortunately,

Canada's Department of Fisheries and Oceans had already done the work. The DFO generously contributed its digital data and basemaps to the MRC, which transferred the information using GIS. In addition, the DFO contributed eight extensive Fish Species Profiles as well as three Ecosystem Community Profiles. These, too, will be linked to the GIS at the MRC office and will be an invaluable source of information for mapping applications.

Another contributor is the Nova Scotia Geomatics Centre, a provincial government agency that not only provides technical assistance to the MRC but also has donated 23 digital files of the new Coastal Series of maps for the Bay of Fundy. These maps, combining data from both topographical maps and navigational charts, are the first of their kind in Canada and will be of considerable importance to both marine-based and coastal activities.

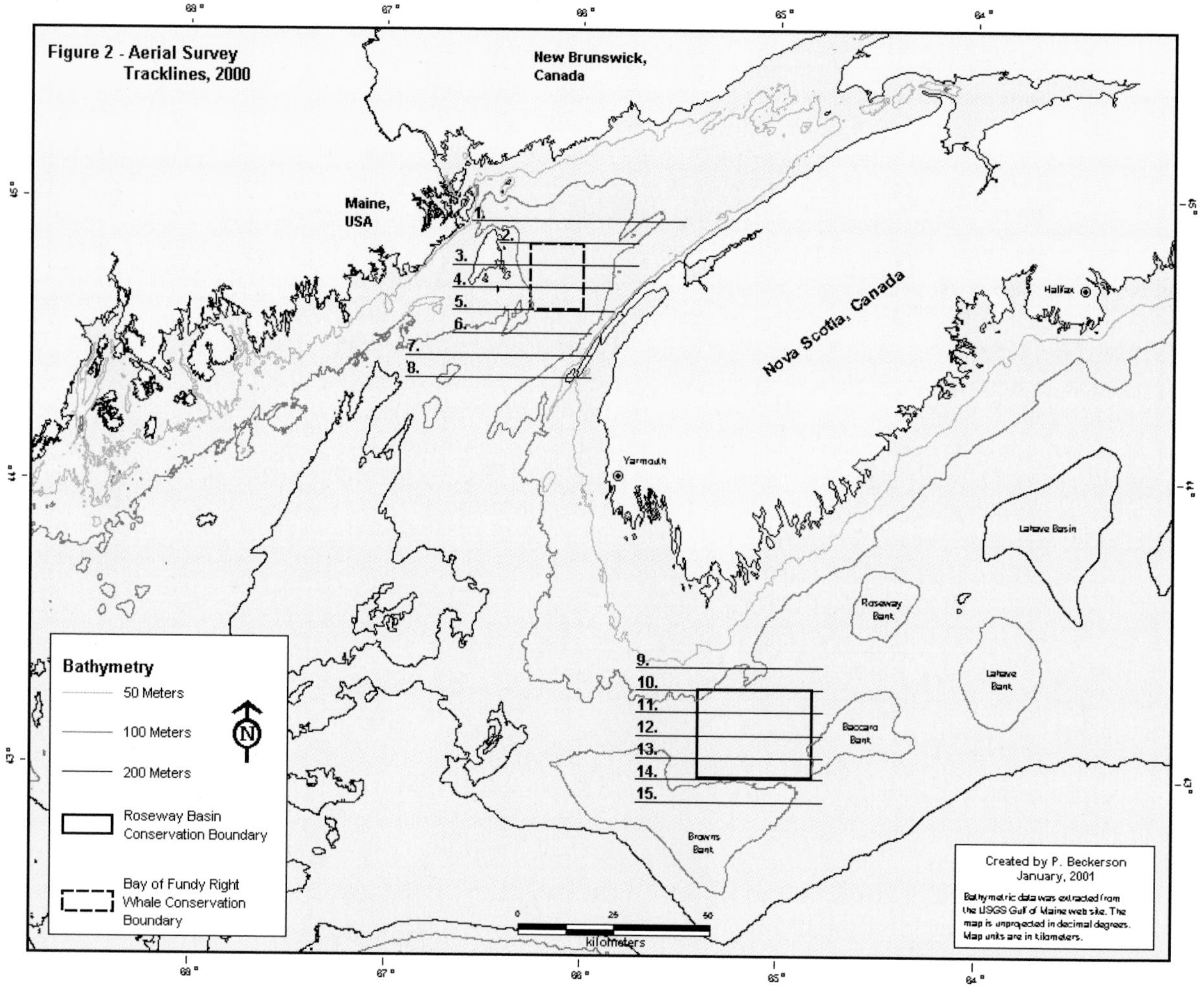

Bathymetric contour lines represent the seafloor depth surrounding the Roseway Basin and the Bay of Fundy and right whale conservation boundaries.

Because of its location on the Gulf of Maine, the MRC has also been very active in working with other community organizations in neighboring New England. In particular, the Cobscook Bay Resource Center in Eastport, Maine, has been a key partner in this cross-border networking effort. Over the last few years, its development has paralleled that of the MRC and, therefore, the two resource centers have come to rely on each other. Among their similarities is a major commitment to bring GIS capacity to marine issues at a grassroots community level. A key part of their cooperation has been to develop innovative ways for facilitating cross-border data sharing. The need for data sharing became apparent as soon as the two centers saw each other's maps, both of which showed blank areas on their opposite borders. A collaborative project is expected in the near future.

The MRC embarked on another project in early 2000 that was funded through the Canadian Rural Partnership Program, a federal government program. The goal of the project was to educate and promote GIS at a community level that included both local groups and fishermen's organizations. At first,

When organizations like the Bay of Fundy's Marine Resource Centre produce maps, they help to make comparison analyses.

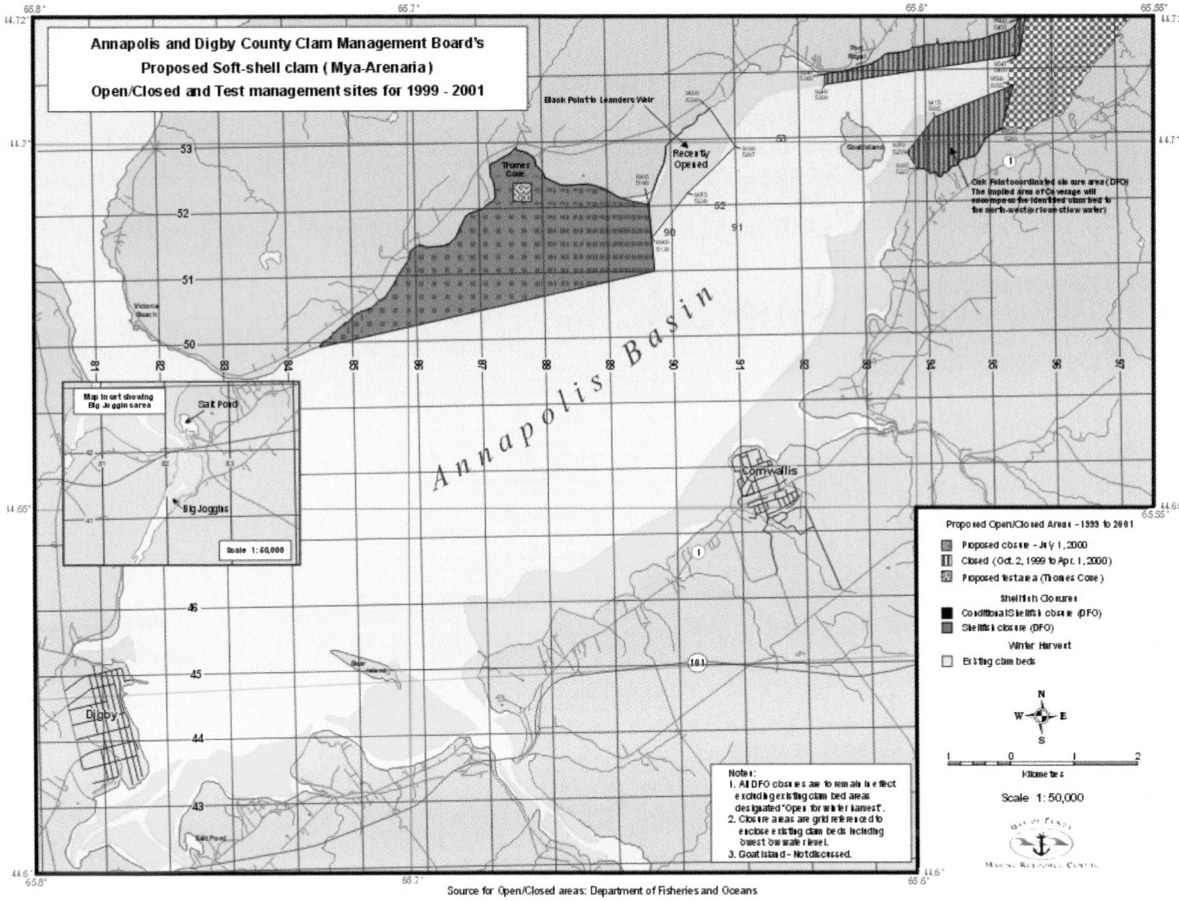

Source for Open/Closed areas: Department of Fisheries and Oceans

some of these groups were not convinced that having a map of their information would be of much immediate benefit to them. However, when agencies were shown the data in a digital format, they began to realize the possibilities available to them with GIS. For example, they saw that GIS has the ability to add layers of data, which is useful for comparative analyses. This function offered the capability to monitor fish catches and their distribution over large areas during a given period of time. Fishermen and regulatory agencies alike found this spatial representation to be of great worth.

When monitoring fish habitats, the GIS Tag Recovery Map serves as a strong analytical tool. The MRC will assist an organization in New Brunswick with the development of a map of the St. John River. The map would show color-coded areas where fish were tagged and then released and also where the fish were then recaptured and the tags recovered. The MRC also provides support to a fishermen's group known as the Fund Fixed Gear Council Ground Fish Project. Together, they are designing a framework for GIS mapping as a component of their ongoing research. Factors considered in the design are issues of data sharing and digital file licenses.

MRC's GIS supports report writing on marine life. The East Coast Ecosystem completes aerial and boat surveys every summer to determine the location of right whales, dolphins, sharks, and other marine mammals found in the Bay of Fundy, particularly around Grand Manan Island and Roseway Bank. In collaboration with worldwide conservation organizations, the primary focus of this work is to track the right whale.

The Bay of Fundy and waters between New Brunswick and Nova Scotia, the proposed region for dragging.

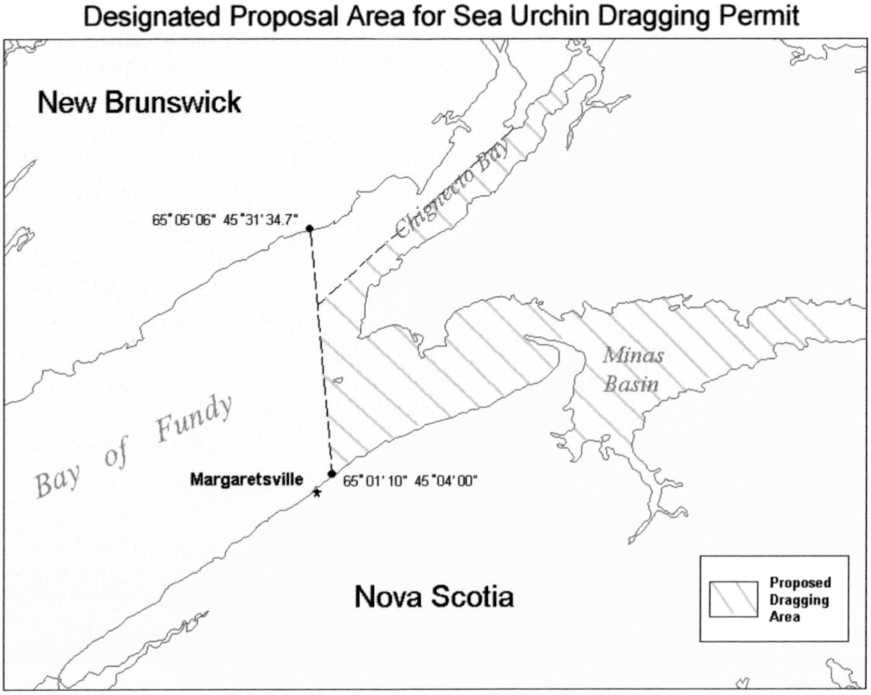

Clams are also important to the Annapolis Basin. The GIS-produced clam map is used to display the current status of particular clam beds during 2000–2001. This map has been used to assist the management board in its decision regarding harvesting practices and conservation issues.

Currently, scientist Sabrina Sturman has begun a suitability study of potential lobster locations in St. Mary's Bay. Her Lobster Larvae Settling Study will not only provide an update of previous data on the crustacean but will also serve to increase the detail of coastal zone maps of that region.

For a presentation at the annual sea urchin harvesters meeting, the MRC prepared five maps of the Digby Gut, Grand Passage, Petite Passage, and a section of coast around Delap's Cove. Based on data provided by the Nova Scotia Department of Fisheries, these maps depicted bathymetry and license condition prohibitions. Smaller-scale maps displaying regional overviews were also generated for the conference. Urchin harvesters will use these maps as working tools to sketch topics of discussion at their meetings, and analysts will use the maps as tools to develop a management plan.

Digby will also use GIS applications for tourism. A 500-acre site, located on the bay's neck, is being proposed for developing an ecoresort. An important component of the project's proposal includes maps of trails, infrastructure, and resources. Other controversial sites include the areas' lighthouses. The Lighthouse Preservation Society looks to MRC for a GIS Web-based map for Annapolis and Digby Counties. This is an essential step in promoting community involvement in stewardship of lighthouses before these marvelous edifices are all sold into private hands.

One more application for GIS is charting abandoned fish weirs. The Western Nova Scotia Yacht Club contributed a chart to MRC of areas in the Annapolis Basin that are affected by abandoned fish weir poles. At low tide, these poles are conspicuous at approximately three feet above the water. However, at high tide or during times of low visibility, the posts are a serious unseen hazard to boaters. The Yacht Club wants to maintain the safety reputation of Annapolis Basin and Nova Scotia for the visiting yacht trade. Therefore, to prevent collision damage, the club seeks to have the fish weirs removed. Based on the information provided by the club's chart, the MRC created a GIS map to show these dangerous sites, a crucial tool for the club when it lobbies to have the hazards removed.

Finally, the MRC's GIS is important in supporting the region's ecological disaster plan. The Canadian Coast Guard is designing an oil spill response plan for the Bay of Fundy. The MRC will head the effort to complete a plan for its side of the bay. This would involve meeting with community leaders, community organizations, and businesses to map out the resources and staging areas for emergency response teams.

The Marine Resource Centre is diligently moving forward in creating a variety of opportunities offered by its GIS. With a strong GIS support team, the center's efforts and foresight have been well placed as it provides the facilities for meeting the challenges of the Bay of Fundy area by designing viable GIS solutions for a multitude of users.

Growing a National Database

Jack Sobel
The Ocean Conservancy
Washington, D.C.

The Ocean Conservancy
Advocates for Wild, Healthy Oceans

The Ocean Conservancy, headquartered in Washington, D.C., has deployed GIS to compile and utilize databases on marine and coastal protected areas (MACPAs) throughout the United States. The Conservancy's initial GIS efforts were directed toward developing the first complete database of federal MACPAs and eventually expanded to include state and regional information. The Conservancy's MACPA database contains standard information including size, year established, and management entity for all sites and more detailed and expanded information for other sites. The Conservancy's GIS work is intended to support public decision making for marine and coastal protected areas and marine conservation.

PHOTO COURTESY OF DON DEMARIA AND WOODFIN CAMP

Redtail triggerfish, an example of marine wildlife protected via creation of the Tortugas Ecological Reserve in the Florida Keys National Marine Sanctuary.

Comprehensive information about marine and coastal protected areas has traditionally been difficult for the public and decision makers to access because it has been scattered among a host of management agencies and institutions. These groups expressed a need for expansive geographical information. Obtaining its initial support from the U.S. Environmental Protection Agency (EPA), the Conservancy initiated efforts to compile a comprehensive database of marine and coastal protected areas in the United States. Over the past several years, the Conservancy has collaborated with its partners to compile related regional and local information. The Conservancy database was developed to allow government agencies, nongovernmental institutions, and the broader public better access to information about protected areas.

A variety of methods have been used for compiling databases. The initial federal MACPA database was created by integrating information from various sources including the EPA, the National Oceanic and Atmospheric Administration (NOAA), the National Park Service, and the U.S. Fish and Wildlife Service. Some important attribute data was also available on federal agency Web sites and in federal agency publications. Good sources for data gatherers were lists of coordinates found in various federal regulations. First

PHOTO COURTESY OF DON DEMARIA (COPYRIGHT)

PHOTO COURTESY OF DON DEMARIA (COPYRIGHT)

PHOTO COURTESY OF DON DEMARIA (COPYRIGHT)

Top to bottom: sea turtle, dolphin, and marbled grouper at Riley's Hump in the Florida Keys.

these were entered into Microsoft Excel, then combined with ArcView, and finally assigned a list of attributes. These attributes included the protected area's name, state, designation, acreage, and the year it was established.

When available, GIS data was collected from the various state and federal agencies. ArcView was used to reproject the data into a common coordinate system. When GIS data was not available, coordinates derived from federal regulations were used to create points and polygons. Coordinates were entered into Microsoft Excel and saved as a DBF file.

The DBF files were then imported into ArcView, and the Add Event Theme feature was used to create the necessary polygons or points. As needed, polygons of protected areas were clipped to a common shoreline to create a clean coast. The various attributes that came with the data were replaced with a common set of attributes. Attributes were obtained from phone interviews, regulations, Web pages, brochures, and other sources. When available, official information about size was used. In some cases, attribute information was calculated using GIS. For example, ArcView was used to calculate the size of some fishery closures that were created from boundary coordinates.

The initial federal GIS MACPA database was put on both CD–ROM and a set of large-format, hard-copy maps, which were delivered to the EPA. Both the Conservancy and EPA have used these resources for a variety of purposes and have posted MACPA maps from them on their respective Web sites. The Conservancy itself has used this information effectively in meetings with government agencies, conservation groups, and the public. As part of its marine protected area initiative, the federal government (NOAA) has developed its own MPA Web site and is developing a more elaborate MPA database. The Conservancy has met with and shared information from its database with NOAA representatives.

The Conservancy has also produced GIS maps of marine protected areas in California from a database compiled by the California Sea Grant. Conservation representatives were able to provide maps to members of the California state legislature that showed the actual amount of marine area in which fishing is completely prohibited compared to the areas considered marine "protected" areas. The maps served as effective visual tools in the Conservancy's ongoing efforts to establish more no-take marine reserves along the California coast.

The Conservancy, in collaboration with several partners, is using the Gulf of Maine region as a prototype for developing improved regional GIS databases on MACPAs. Data in the Gulf of Maine includes state protected areas and indicators of the extent of protection offered by each area. The Conservancy used telephone interviews, regulations, and other information sources to obtain information on management objectives and regulatory measures that address pollution, and protection of benthic habitats, species, and coastal habitats. The Conservancy also convened a panel of regional experts to develop a model that derives a conservation value based on the size, permanence, and seasonality of each area. The results of all three analyses are contained in the expanded database, which the Conservancy hopes can be modeled in other priority areas. Information from the expanded database is currently being used to develop a book and poster that will be an important tool for the Conservancy staff in New England.

The Conservancy uses ArcView and ArcView Spatial Analyst software to integrate information from its initial marine protected area (MPA) and MACPA GIS database efforts with information from other sources to help evaluate existing and proposed MPAs and develop more effective and comprehensive MPA networks for the future. These integration efforts have involved physical information (e.g., bathymetry and currents), biological information (e.g., species and habitat distribution), and human-use information (e.g., fishing effort and diving locations). The Conservancy is also increasingly integrating GIS into its marine debris, fisheries, pollution, and protected species work.

Author Jack Sobel on a research expedition to help create the Tortugas Ecological Reserve in the Florida Keys National Marine Sanctuary.

PHOTO COURTESY OF DON KINCAID

PHOTO COURTESY OF DON DEMARIA (COPYRIGHT)

Hogfish, Riley's Hump.

Pacific

Farming Coral Reef Invertebrates for Reef Rehabilitation and the Aquarium Trade

INTERNATIONAL MARINELIFE ALLIANCE

Peter J. Rubec
International Marinelife Alliance
St. Petersburg, Florida

Joyce Palacol
International Marinelife Alliance
Pasig, Philippines

Coral reefs and other coastal habitats throughout Southeast Asia are being destroyed by harmful fishing methods (Rubec 1988). Explosives are used to kill fish for human consumption. Scare-lines (ropes with streamers attached to large rocks) are used to pound corals in order to drive food fish out of the reef into nearby nets. Likewise, kayakas fishing involves poles used to smash corals and drive fish into nets. Illegal trawling over fragile bottom habitats is yet another practice that is reducing sustainable yields to food fisheries and the aquarium trade.

Divers squirt cyanide solution (sodium cyanide tablets dissolved in plastic detergent bottles) onto coral heads to capture live aquarium and food fish seeking refuge in the reef (Rubec 1986, 1988; Rubec et al. 2001a). Cyanide doses, exceeding several thousand parts per million, kill about 50 percent of the exposed fish. The others appear to recover if quickly moved to clean water. About 80 percent of the live marine aquarium fish exported from the Philippines, Indonesia, and Vietnam die during transport to consuming nations from the detrimental affects of cyanide, ammonia, and stress. Similar delayed mortalities occur with live groupers caught using cyanide to supply gourmet restaurants in Hong Kong and mainland China (Johannes and Riepen 1995). Fishermen also scatter cyanide tablets over the reef to kill food fish sold in local markets.

Early reports that cyanide kills coral reefs (Rubec 1986) have been confirmed by controlled experiments (Cervino et al., in press). Hard and soft corals (genera *Scolymia, Goniopora, Euphyllia, Acropora, Heliofungia, Plerogyra, Favia, Sarcophyton*) exposed for three minutes to cyanide concentrations ranging from 50 to 600 parts per million exhibited loss of color associated with the expulsion of zooxanthellae (bleaching), disruption of protein synthesis, and altered rates of cell division. Mantle tissue died and peeled off the skeletons with most genera tested, and the majority of test specimens died within two months.

The Philippines has laws against destructive fishing. Unfortunately, the laws are difficult to enforce because there are more than 770,000 small-scale municipal and commercial fishermen and more than 4,000 aquarium-fish collectors spread across 7,000 islands. Various levels of government need the ability to collect and record accurate environmental and fisheries data, analyze the information geographically, and produce management recommendations in a timely manner. Therefore, it is necessary to balance the protection of marine resources against the enhancement of fisheries and mariculture production. There is an urgent need to turn fishermen toward other occupations through alternative livelihood training programs.

The International Marinelife Alliance (IMA) is a nonprofit, nongovernment marine conservation organization with offices in the United States, Philippines, Indonesia, Vietnam, Hong Kong, Guam, Fiji, Vanuatu, Marshall Islands, Federated States of Micronesia, and Australia. The goals of IMA are to conserve marine biodiversity, protect marine environments, and promote sustainable use of marine resources by local people.

Until 1998, the community-based programs conducted by IMA consisted of training cyanide fishermen to use less-destructive fine-mesh barrier nets to capture marine-aquarium fish and to use hook-and-line methods to capture live food fish such as groupers and snappers for export to Hong Kong, Taiwan, and mainland China (Rubec et al. 2001a). However, with the publication of recent scientific studies demonstrating that the use of cyanide destroys coral reefs, the IMA has increased emphasis on mapping coastal habitats by using geographic information systems (GIS) to support habitat conservation.

The training programs require coordination with municipal governments and other nongovernment organizations. When introducing the Destructive Fishing Reform Program (DFRP) to an area, the IMA training team discusses the initiative first with local village officials and then with the fishermen. They also discuss ways to increase income either from fishing or other alternative livelihoods. The sustainable use of natural resources is addressed through the community-education component of the DFRP (Rubec et al. 2001a). Local management options are considered such as limiting access, policy enforcement, or sanctuary development.

IMA staff provided assistance and GIS training to personnel in the Philippine Bureau of Fisheries and Aquatic Resources (BFAR) in the central office, field offices, and also selected municipalities situated in the provinces of Cebu, Bohol, and Palawan. The training provides students with a working knowledge of GIS concepts and applications so that they may assist these government agencies with the management of the country's coastal resources. Part of the GIS training has focused on the determination of boundaries using ArcView software to define the spatial extent of municipal waters.

After GIS training, the next step is the development of the database. Five sites in three provinces were chosen for the GIS mapping: Olango Island and Lapu Lapu City on the Island of Cebu, Guindacpan and Talibon on the Island of Bohol, and Santa Cruz in Davao del Sur, Southern Mindanao. The IMA's GIS personnel have created maps for 11 villages (barangays) on Olango Island situated in the municipality of Lapu Lapu near the Island of Cebu, two villages in northeastern Bohol, and three villages situated in the Davao Gulf of Davao del Sur.

The GIS work was geared to the creation of coverages to support community decision making such as deciding where to establish a marine protected area, and deciding the areas most suitable for various types of mariculture. The maps for each village show coastline areas and depict terrestrial and marine zoning, bathymetry, ocean currents, salinity, and temperature. Symbols used with the composite habitat maps indicate areas where different fish and invertebrate species are caught; the location of stationary fishing gears; areas with destructive fishing; the boundaries for marine protected areas and wildlife sanctuaries; human-use areas such as public parks, resorts, and marinas; and locations with coastal pollution *(figure 1).*

The composite habitat maps depicting terrestrial and marine zones such as the types of shoreline, mud flats, sea-grass beds, passes, channels, reef flats, and barrier reefs were created from multispectral SPOT® satellite imagery (Système Probatoire d'Observation de la Terre) obtained from the Philippine National Mapping and Resource Information Authority. The scale of the rectified maps created using ArcInfo and ArcView ranged from 1:8,500 to 1:11,500. Bathymetry contours were digitized from nautical charts. Seasonal salinity and temperature data was interpolated using ArcView Spatial Analyst from point measurements gathered by BFAR staff using data loggers. The fisheries species distribution data was obtained from interviews with local fishermen.

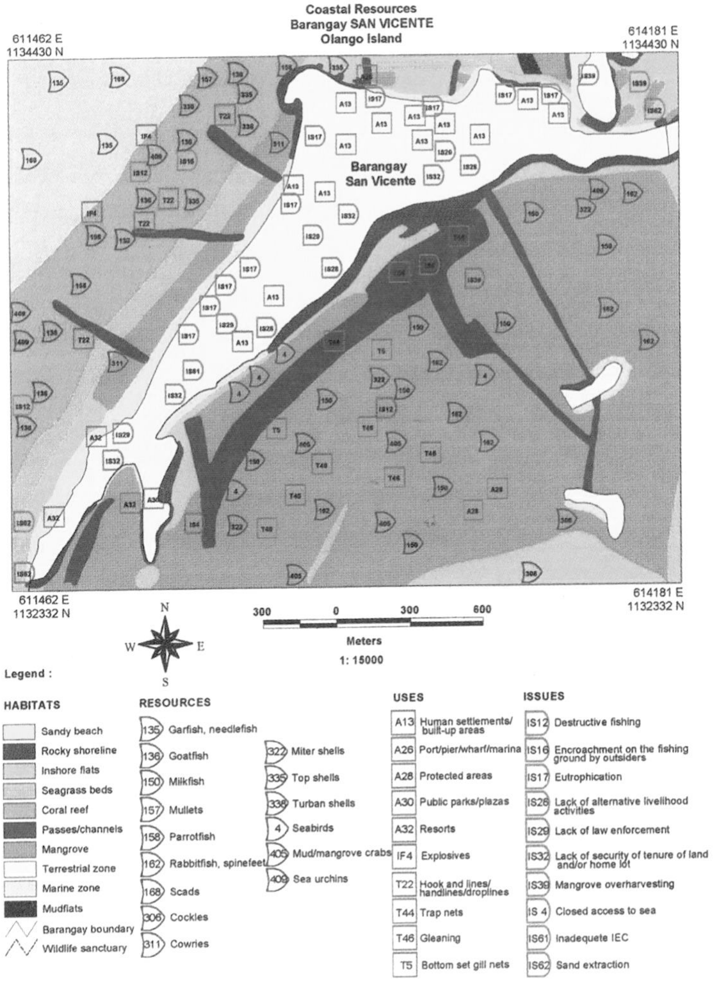

Figure 1: Marine resources map of Barangay San Vicente on Olango Island depicting locations of various fish species and the location of various stationary fishing gears and other human-use areas, in relation to benthic habitats mapped using SPOT remote-sensing imagery.

A promising means by which community-based coastal resource management programs can be implemented in the Philippines is through territorial use rights in fisheries (TURFs). Community control of the means of production through TURF management has the potential of resolving user conflicts and reducing fishing effort in specific areas (Rubec et al. 2001b). Changes in the Local Government Code in 1991 and the new Fisheries Act in 1998 granted municipalities the right to license fisheries and lease mariculture sites within municipal waters. Currently, municipal governments through Fisheries and

Aquatic Resource Management Councils (FARMCs) have the authority to regulate the implementation of TURFs. The potential of using GIS associated with environmental monitoring and spatial management strategies such as zoning areas for sanctuaries, for certain fisheries, or as sites for mariculture are now being evaluated.

Corals and other reef-invertebrate species reared on TURFs for export could become an important source of revenue for local communities (Rubec et al. 2001b). It may be possible to halt the destruction of coral reefs by demonstrating that reef fish and invertebrates can be reared profitably for the aquarium trade. This should be tied to programs to convert small-scale fishermen to TURF farmers.

The IMA has been using GIS to delineate the boundaries of marine protected areas and TURFs for mariculture through consultation with the municipalities, FARMCs, BFAR, and the fishermen. The GIS database is being used to assist with this planning process.

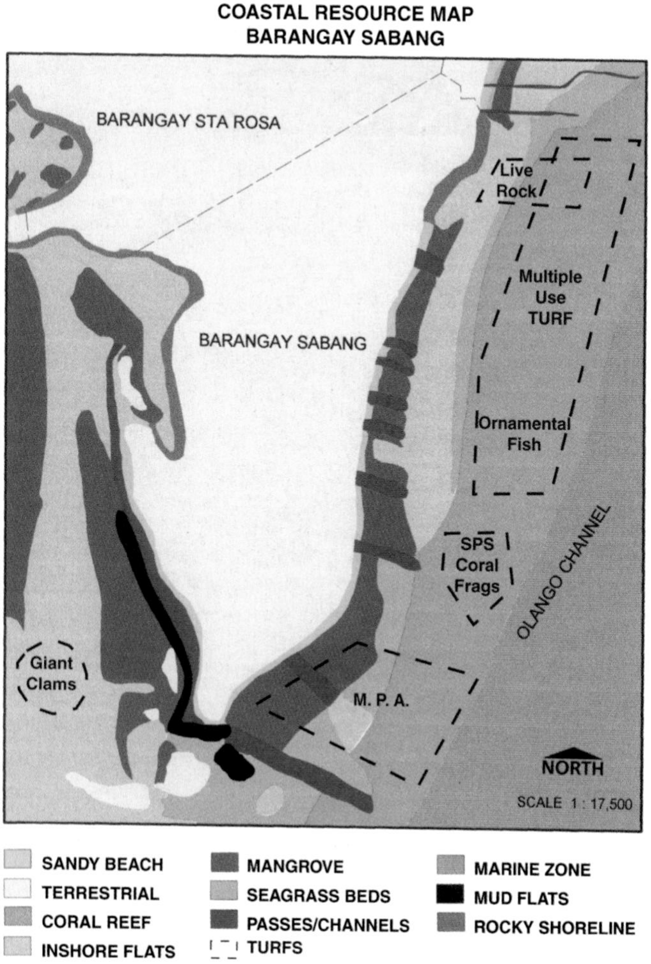

Figure 2: Marine resource map of Barangay Sabang on Olango Island depicting hypothetical locations of a marine protected area (MPA), and territorial use rights in fisheries (TURF) areas designated for collecting ornamental fish and the culture of giant clams, coral fragments, and live rock.

The first step of the planning process is to create habitat maps that show species distributions and then to analyze the maps to determine which areas are most suitable for mariculture. Areas already being used for fishing are excluded from consideration; likewise, areas with a high level of silt, or otherwise unfavorable water quality from pollution, are also excluded. Next, underwater surveys are conducted to determine which of the remaining locations would best support the farming of giant clams, coral fragments, or live rock. Finally, the mariculture potentials of the sites are rated. Sites close to existing coral reefs are rated more highly, since they already support populations of the organisms desirable for mariculture. Figure 2 depicts the location of hypothetical TURFs overlaid onto the habitat maps in the village of Sabang on Olango Island.

The IMA is working in cooperation with government organizations and the FARMCs to plan the placement of TURFs. Key habitats such as mangroves, sea grass, and coral reefs will be protected, while other areas are leased to members of the community. The goal is to shift fishermen away from destructive fishing by demonstrating the economic benefits of farming reef organisms, while raising community awareness of the need for conservation of coastal habitats.

GIS-produced maps support decision making concerning the allocation of lease sites used as TURFs. The FARMCs can limit access by mobile fishermen using destructive fishing methods. Members of the community are able to police the TURFs because they have control over the resources being farmed. Hence, GIS becomes a key tool to support spatial management of these vital marine resources.

Hard and soft coral fragments

In an effort to rehabilitate damaged reefs, scientists have developed techniques for transplanting corals and culturing fragments attached to artificial substrates. The coral fragments can be propagated in TURFs situated on reef flats.

A coral farm is already well established on Olango Island. James Heeger of the University of San Carlos (situated in Cebu City on the island of Cebu) created a coral farm in the village of Caw-oy. An environmental training center was established to train local fishermen in the basic skills of coral farming. This includes methods for selecting donor corals, applying fragmentation techniques, monitoring and maintenance during the grow-out period, and setting up cooperatives for marketing the corals. The fishermen place coral fragments in concrete frames deployed on the bottom in TURFs selected by the community *(figures 3 and 4).*

The IMA recently took over the management of the Caw-oy coral farm, which has become a regional training center. The IMA has initiated projects to rehabilitate coral reef areas and expand training in coral farming to other villages on Olango Island, on the island of Bohol, and in the Davao Gulf situated in southern Mindanao. Using corals from Caw-oy, the IMA has created a second coral farm to support reef rehabilitation at the village of Consuelo on Camotes Island off Cebu. Moreover, the IMA is also promoting ecotourism at the coral farm by creating underwater trails for divers.

PHOTO COURTESY OF PETER J. RUBEC

PHOTO COURTESY OF PETER J. RUBEC

PHOTO COURTESY OF PETER J. RUBEC

Figure 3: Two divers breathing air through plastic hoses from a Brownie Third Lung hookah apparatus in a rubber raft, situated in front of the platform (top); a diver placing the concrete frame over plastic sheeting on the bottom. The plastic is used to protect the coral fragments from predatory snails and crustaceans in the sand (center); divers breathing air through hookah hoses while placing coral fragments in the concrete frames spaced over the bottom (bottom).

PHOTO COURTESY OF PETER J. RUBEC

PHOTO COURTESY OF PETER J. RUBEC

PHOTO COURTESY OF PETER J. RUBEC

Figure 4: The coral farm at the village of Caw-oy on Olango Island: A platform associated with the coral farm, situated near the village (top); A square concrete frame is lowered into the sea (center); soft coral *(Sarcophyton)* attached to a concrete slab prior to placement of the coral fragments on the seafloor (bottom).

Giant clams

Generally, it takes three to five years to grow giant clams large enough (three to five inches) to harvest for human consumption. In contrast, it takes less than one year to rear giant clams to a length of about one inch for export to the aquarium trade. Hence, the export of giant clams to the aquarium trade can provide more immediate economic returns than harvesting them for food.

The IMA has established a sanctuary for holding giant clam brood stock. A hatchery for giant clams at the University of the Philippines is producing clam spat (animals less than $1/16$ of an inch in size) for distribution to coastal communities. About five thousand giant clams obtained from the university presently are being reared on TURFs situated near the village of Caw-oy *(figure 5)*. Fishermen will be trained to rear the clams in cages situated on TURFs allocated by the FARMCs. The giant clams will be used for restoration of wild populations and for export to the aquarium trade.

Live rock

The IMA plans to train Filipino fishermen to create artificial live rock from coral sand mixed with concrete molded into various forms. The artificial rock will be deployed on TURFs in coral reef areas, where the rocks can become coated with coralline algae and then be colonized by planktonic larvae produced by marine invertebrates (sea anemones, zoanthids, crinoids) situated on nearby reefs. After about a year, farmers will be able to harvest the live rock for export to the marine-ornamental trade.

The IMA is assisting the communities to obtain CITES (United Nations Convention on International Trade in Endangered Species of Flora and Fauna) export permits from the Philippine government. These permits will allow coral fragments, giant clams, and live rock reared on TURFs to be exported to the aquarium trade to enhance the income of local people.

For TURF farming to become a reality, various integrated strategies need to be developed, including village-level training and loan programs that assist local communities to create hatcheries and other infrastructure. Scientists need to become more involved in training programs to transfer their knowledge to fisher folk. The use of GIS can assist with zoning coastal near-shore areas in a manner similar to terrestrial coastal planning.

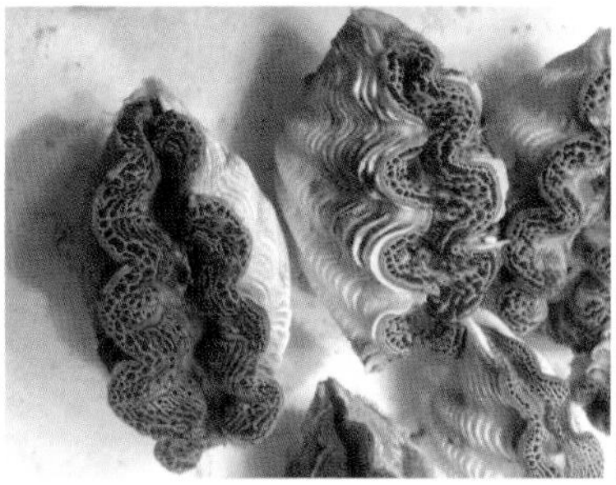

Figure 5: Giant clams from the University of the Philippines are reared at the coral farm site located near the village of Caw-oy on Olango Island.

References

Cervino, J. M., R. L. Hayes, M. Honovitch, T. J. Goreau, S. Jones, and P. J. Rubec. Changes in zooxanthellae density, morphology, and mitotic index in hermatypic corals and anemones exposed to cyanide. *Marine Pollution Bulletin,* in press.

Johannes, R. E., and M. Riepen. 1995. *Environmental, economic, and social implications of the live fish trade in Asia and the Western Pacific.* Honolulu: Nature Conservancy Report.

Rubec, P. J. 1986. The effects of sodium cyanide on coral reefs and marine fish in the Philippines. In *Proceedings, The First Asian Fisheries Forum, Manila, Philippines: Asian Fisheries Society,* edited by J. L. Maclean, L. B. Dizon, and L. V. Hosillos, 297–302.

Rubec, P. J. 1988. The need for conservation and management of Philippine coral reefs. *Environmental Biology of Fishes* 23(1–2), 141–54.

Rubec, P. J., F. Cruz, V. Pratt, R. Oellers, B. McCullough, and F. Lallo. 2001a. Cyanide-free net-caught fish for the marine aquarium trade. *Aquarium Sciences and Conservation 3*:37–51.

Rubec, P. J., V. R. Pratt, and F. Cruz. 2001b. Territorial use rights in fisheries to manage areas for farming coral reef fish and invertebrates for the aquarium trade. *Aquarium Sciences and Conservation* 3:119–34.

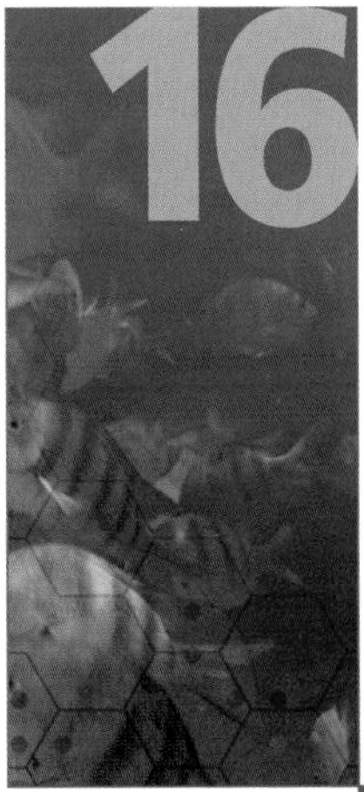

16

Heal the Bay Community Action Group Uses GIS to Change Policy and Inform the Public

Mark Abramson and Damon G. Wing
Heal the Bay
Santa Monica, California

A small group of citizens can indeed fight city hall and make a huge difference in conservation policy. Citizens of the Malibu Creek Watershed area started with a vision of restoring their watershed and took action that has resulted in statewide changes. They continue to thrive and serve as a model for volunteerism, technological application of GIS, chemical water quality testing, and political advocacy for many conservation groups.

The Malibu Creek Watershed encompasses 110 square miles of some of the most scenic natural and recreational resources in Southern California. This watershed is home to many different types of land uses ranging from the rural undeveloped Santa Monica Mountains to dense urban developments. These natural and recreational resources draw millions of visitors every year.

Three of the most recognized resources within the watershed are Malibu Surfrider Beach, Malibu Lagoon State Park, and Malibu Creek State Park. These sites serve as recreational areas for beachgoers, birdwatchers, hikers, mountain bikers, naturalists, surfers, and swimmers. This beautiful area is also home to numerous species of wildlife including the endangered California gnatcatcher, steelhead trout, red-legged frog, tidewater goby, and the California brown pelican. The problem is that as development continues to grow, so does urban runoff that negatively influences the ecology of the watershed and endangers the health of people who recreate there.

In the past, Los Angeles County treated its beaches and coastal waters as dumpsites. Because of rising concerns about the habitat, a small group of citizens took it upon themselves to heal this ecologically troubled area that was suffering from some of the worst levels of contamination found anywhere on the nation's coastlines. This group of advocates solicited the help of the California Coastal Conservancy

The Stream Team uses 25-foot telescoping antennae poles to rise above the tree canopy and acquire satellite signals to perform their surveys. The GPS team will look for discharge points and outfalls that could be sources of creek pollution.

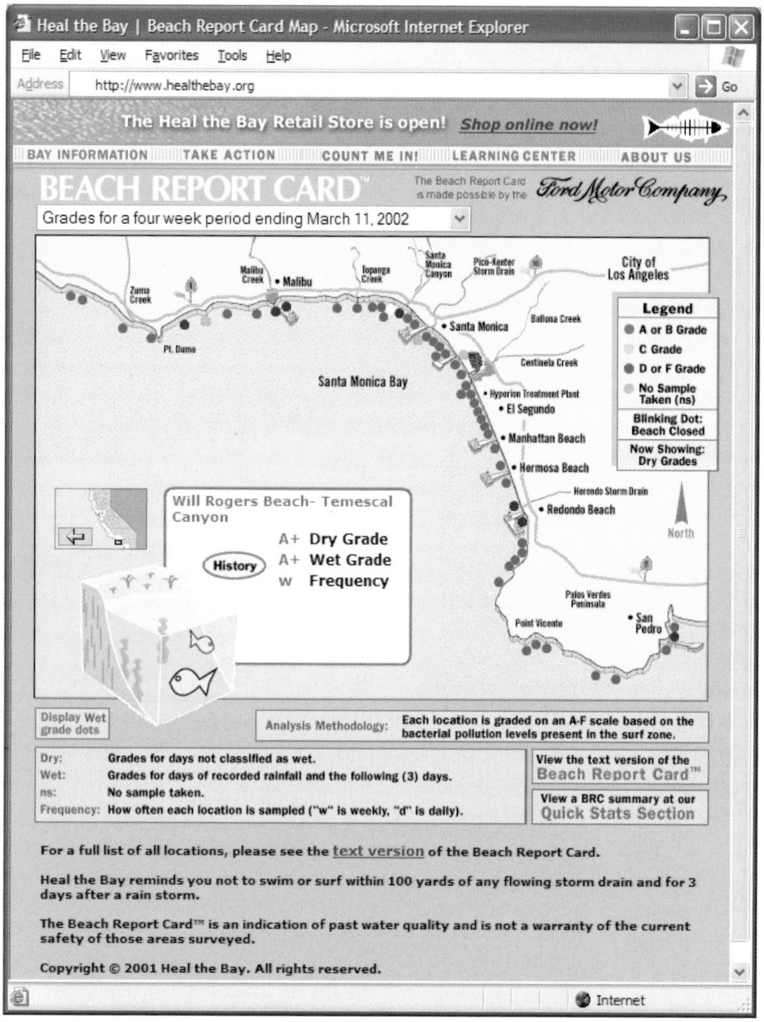

Heal the Bay's Web site features a Beach Report Card that rates water quality at local beaches.

to initiate a conservation project that was soon dubbed the Stream Team. A supportive conservation grant was also issued by ESRI providing GIS software, training, and technical support. Volunteer citizens were eager to help in the effort to gather and record information within the Malibu Creek Watershed.

Heal the Bay contracted with the California State Polytechnic University, Pomona, graduate program, Department of Landscape Architecture, to map and model the natural processes of the Malibu Creek Watershed. These students helped design the Stream Team Program, including a method to enter data collected at precise locations using ArcView software. The Stream Team was designed as a three-pronged approach to monitor water quality, map stream habitat, and measure the biological integrity throughout the Malibu Creek Watershed. Volunteers were trained to measure water quality, map in-stream and riparian habitats, and collect benthic macroinvertebrate samples to determine the effects of pollution on biological communities. Benthic macroinvertebrates are insects that live primarily on the bottom of streams and rivers from three months to three years of their life cycle. Because these insects are in constant contact with the water, they are good indicators of long-term water quality and help researchers assess the impacts, if any, from poorly implemented developments.

Stream-mapping volunteers are trained and certified to operate Global Positioning System (GPS) technology and to precisely map and conduct stream habitat assessment. Mapped data is ultimately put into ArcView, an integral component of Heal the Bay's GIS. It has been the goal of this program that information collected is useful and useable by the numerous agencies and organizations vested in protecting the resources throughout this watershed. Because the training program has had such great success, other agencies such as the National Park Service, state parks, municipalities, and other volunteer monitoring programs have also been enrolling and applying this knowledge to their own particular conservation efforts.

Stream Team volunteers walk along the streams and creeks in the Malibu Watershed that eventually empty at Malibu Lagoon State Park and world-famous Malibu Surfrider Beach. Combining GIS and chemical-testing techniques, the team locates illegal or illicit connections that flow into the creeks, identifies areas that are environmentally damaged or disturbed, and tests the water quality of different locations within the watershed. By taking digital photographs, entering coordinates, and creating site reports, the team provides a full depiction of at-risk areas.

Those who work in the GIS component use GPS capable of submeter accuracy. The GPS unit is well suited for its task. Sometimes data collection techniques must be creative because of the satellite positioning relationship to difficult areas. For example, the Stream Team uses 25-foot telescoping antennae poles to rise above the tree canopy and acquire satellite signals to perform their surveys in this difficult terrain.

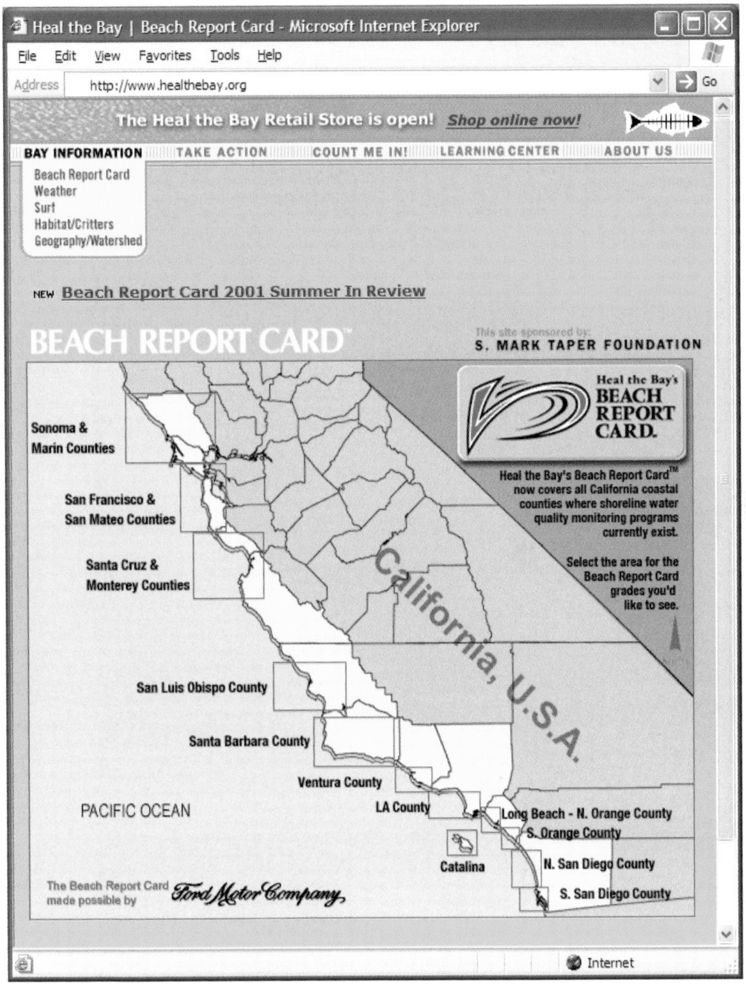

The Beach Report Card covers all California coastal counties where shoreline water-quality monitoring programs currently exist.

The GPS team will look for discharge points and outfalls that could be sources of creek-damaging pollution. These pipes are GPS recorded, and then volunteers complete data sheets about the site incidents. Other events that are recorded into the GIS database include unstable stream banks, exotic invasive vegetation, in-stream pool habitat, land uses that are obviously impacting the stream, dump sites, artificial stream bank modifications such as concrete banks, and impairments within the stream channel. Each time one of these items is mapped, a digital photograph is taken of the exact item. The digital photographs are then hot-linked in ArcView to the exact location of the item mapped. Combining maps and digital imagery allows the data to be more easily understood by decision makers and the general public.

In the top photo, workers collect and record environmental data. The bottom photo shows the damaging effects of boulders placed along the sides of the creek.

Malibu is well known for its landslides, and data collectors make a special effort to map the creek's unstable stream banks. The sedimentation these slides generate creates a big impact on the watershed. For instance, slides affect the steelhead habitat and impede steelhead spawning areas. Frequently, unstable stream banks are correlated with impervious surfaces such as areas that have been covered with concrete. Because these surfaces no longer absorb water, the creek bears faster and larger flows downstream, which in turn scour its outside banks. Developers' efforts to shore up the creek's sides with boulders only exacerbate the problem because this change creates a funnel effect that intensifies the flow downstream. Because of their environmental impact, these fortifications are also mapped.

A wide range of vegetation is part of the creek system that cuts completely through the Santa Monica Mountains. The team documents the watershed's riparian habitat. A research team has most recently discovered the San Fernando Valley spine flower. This indigenous flower, previously assumed extinct, exists only along a limited strand of the watershed. Intrusive exotic vegetation is also entered into the database. Identifying these areas, the GIS spatially indicates where native vegetation is being pushed out by exotic invasive vegetation. The GIS offers analytical tools for targeting exotic plants for removal and marking areas of natural vegetation.

The Stream Team also maps habitat within the stream channel. Here, the creek supports pool habitats that serve as holding places for fish and frogs. A project team discovered a rare red-legged frog in the headwaters of Las Virgenes Creek. This habitat, dutifully plotted into the database, was not only site-referenced but pictorially illustrated as well. This style of presentation makes a strong impact in educating the public. Some project supporters and watershed stakeholders specifically expressed concerns about the distribution of different exotic species along the creek. GIS mapping is capable of establishing trends that provide a history of these movements.

In an effort to protect the receiving water area, the team maps the location of impairments created by algae and sediment to quantify the level of these impairments. Nuisance algae reflect a nutrient issue in the ecosystem. When it reaches an estimate of greater than 30 percent, the area is considered impaired. However, nuisance algae is not really quantifiable, so the team maps it as a line and uses a 10-percent range to estimate the percent of stream channel covered by algae (e.g., 60 percent to 70 percent) to indicate the severity of the impairment. This information is made available along with linked images so that decision makers can see locations and pictures of the extent of the impaired area.

Hot spots indicated by the GIS offer signposts for Heal the Bay chemistry teams to check these areas' pollutant levels. Currently, water chemistry testing is done at 10 locations throughout the Malibu Creek Watershed. The organization trains volunteers in how to collect and test water. A few of the properties tested for are pH, stream flow, dissolved oxygen, nitrates, phosphates, and ammonia. They also test for the entercoccus, total coliform, and *E. coli* bacteria. By combining the GIS database with the chemistry analysis database, water quality trends can be spatially depicted, and areas with consistent problems can be readily identified.

The monthly water chemistry outcomes are published on Heal the Bay's Web site at *www.healthebay.org/StreamTeamData/waterchem.html.* Interested users can visit Heal the Bay's site to download raw water chemistry data free of charge. In addition, the Web interface allows the site visitor to interactively query data by date range, site location, and chemistry parameter. A graph can then be built from the queried data.

As part of the data collection to measure the overall health of the watershed, Heal the Bay will also map impervious surface cover to monitor its impacts on coastal riparian ecosystems throughout the watershed. This study is designed to assess the impacts caused by upstream impervious surfaces in arid climates. The researchers will use ArcView to map and categorize land uses by heads-up digitizing on top of a high-resolution digital orthophoto of the watershed. Areas of unstable streambanks and streambank modifications will be compared against water quality data and benthic macroinvertebrate surveys to determine impacts, if any, to the biological health of streams.

The success of Heal the Bay is found in its influence in public policy making. The mapped data that volunteers gather is forwarded to government agencies responsible for protecting the watershed's resources. Stream Team chemistry data is being used to help develop water quality limits for nutrients in Malibu Creek, and mapping data has been used to plan a large-scale invasive vegetation removal project and monitor its success. Last year, California Governor Gray Davis signed a mandate to allocate $33 million for the improvement of California coastal beaches. At the state level, Heal the Bay helped to draft AB 411, a vital piece of legislation that was implemented as part of the California Health and Safety Code providing the public with "right-to-know" information about water quality at local beaches. This requires that those beaches serving more than fifty thousand visitors a year must conduct weekly testing of three specific bacteria indicators: total coliform, fecal coliform, and entercoccus. Based on the AB 411 data, warning signs must be posted by the Health Department if state standards for bacterial indicators are exceeded. Heal the Bay compiles this data to produce a weekly Beach Report Card that grades over 450 beaches from Sonoma County to the Mexican border A to F.

Users who visit Heal the Bay's Web site (*www.healthebay.org/baymap*) are greeted by a map depicting 12 coastal counties throughout California. Web site visitors can select the county they plan to visit and then further discover the water quality conditions of beaches within that county. Beachgoers for the first time ever can make informed decisions about where they and their families go to swim and recreate.

Now in its second decade, Heal the Bay continues its fight to find workable solutions to the problems threatening the future of the bay and all of Southern California's coastal waters.

17 Effective Use of Volunteer Resources

GIS in Community Activism

Chad Nelsen and Mark Rauscher
Surfrider Foundation
San Clemente, California

Surfrider Foundation
www.surfrider.org

People who live near the coast love the oceans and gladly volunteer to help preserve them. Many conservation organizations are formed as the result of a communal desire to not only initiate cleanup projects, but also gather information and organize educational programs. Originally, the Surfrider Foundation focused on Southern California coastal areas, where it had such success in organizing the efforts of volunteers that the group quickly expanded to 50 chapters throughout the nation.

The Surfrider Foundation's core competency is community-based education and activism. To strengthen and build on this grassroots educational focus, the organization must facilitate the dissemination of up-to-date, science-based information at the community level. Surfrider has been able to accomplish this by the effective development of programs such as Beachscape, its popular community-based coastal mapping program.

Using Beachscape, volunteers document the physical characteristics, land-use patterns, pollution sources, public access, erosion, habitat, and wave characteristics of the nation's coastlines. The aim of this program is to mobilize the Surfrider Foundation's vast national network of local chapters and volunteers to characterize local coastal areas at a scale smaller than is currently available from most data sets.

Volunteer mapping empowers caring citizens.

The Surfrider Foundation's members represent an enormous workforce of interested citizens who have knowledge of their local community. Tapping this community resource enables the Beachscape program to develop data sets at a scale that would be prohibitively expensive for traditional, contract-based data collection projects.

The project uses geographic information systems to store, analyze, and publish data about the influence on coastlines by both natural and human-influenced conditions such as the locations of outfall pipes, hard structures, erosion "hot spots," accumulation of marine debris, and beach accesses. The Beachscape program illustrates cumulative impacts and provides a basis for evaluation of coastal projects and management proposals. Thus, this program

The shoreline of Laguna Beach, California, is heavily impacted by dirty water coming from stormwater drainage.

empowers local citizens with the information and skills they need to be effective advocates for coastal resource protection in their communities. Beachscape is implemented through the Surfrider Foundation's national network of chapters, whose members collect existing GIS data and nondigital maps of beach features, and volunteer to field map the beach. Beachscape has three levels of implementation: Basic, Intermediate, and Advanced.

Extensive shoreline structures protecting private homes reduce the width of the public beach.

Basic Beachscape

Basic Beachscape is the project's simplest level of beach mapping. Basic Beachscape involves using a one-page form to map a local beach. Accompanying the form is a mapping guide that explains how to fill it out. Available to all chapters, Basic Beachscape is an "entry-level" activity that is relatively simple to implement. Team members can enter the collected information into a GIS database as well as keep it in a binder at the chapter level.

The methodology for Basic Beachscape is relatively simple. Using a U.S. Geological Survey (USGS) 7.5-minute topographic quadrangle (topo quad), the chapter members delineate their coastal area into discrete beach sections. These sections are numbered and named. Volunteers are then sent to the beach to collect beach attribute information using the Basic Beachscape form. The information collected is then entered into a Microsoft Access database. A beach delineation theme is generated in ArcView by creating line segments from a coastline derived from the 7.5-minute USGS quadrangle. The attribute information can then be linked to the beach segments in ArcView.

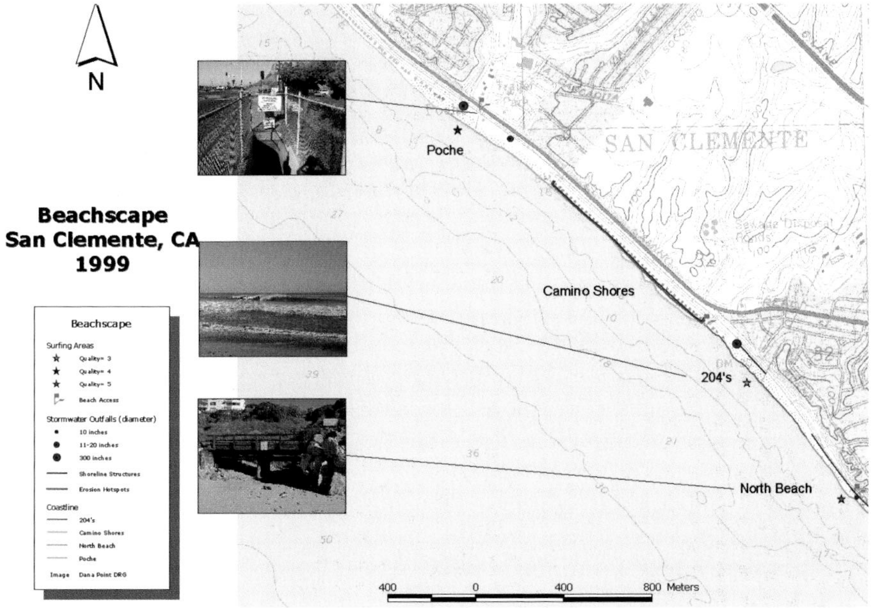

Various attributes of a beach are mapped using Beachscape protocols.

Intermediate Beachscape

Intermediate Beachscape is a more complex mapping program. It involves mapping specific features on a USGS topo quad and filling out attribute forms specific to the feature. This project requires a Beachscape coordinator at the local chapter level and entails some training of the coordinator and users. The collected information is entered into a database and digitized in a GIS. The data is made available via compact disc, paper maps, and the World Wide Web.

The methodology for Intermediate Beachscape involves field-mapping conventions, database design, heads-up digitizing, and project and layout development in ArcView. Field mapping involves marking the locations of specific beach features on a USGS topo quad using a standardized mapping protocol. The team records feature attribute data by filling out a paper feature attribute form and photographing the feature.

If the chapter has a differential Global Positioning System unit, this is used to map the feature more accurately. Once the fieldwork is completed, the data is reviewed for an assessment of accuracy and quality by the chapter coordinator. The maps and data forms are then sent to the Surfrider Foundation national office to be entered into the GIS. The attribute data is entered into a Microsoft Access relational database. Next, the feature data from the maps is heads-up digitized using the USGS digital raster graphic (DRG) of the topo quad. A separate theme is created for each beach feature.

Finally, the database is imported into ArcView and linked to the features in the theme. The photographs are also included and can be linked using the hot link. ArcView layouts are also created to illustrate a basemap for the beach. This information is then made available to the chapter on compact disc, printed maps, and over the Web. The Surfrider Foundation intends to make all data collected for Beachscape publicly available.

Advanced Beachscape

Advanced Beachscape, still in the prototype phase, is the third tier of Beachscape. It is the most complex phase and requires an increased level of commitment from associates. Advanced Beachscape involves Beach Stewards who will "adopt" a stretch of coast and map it seasonally or even monthly. These Beach Stewards' mapping efforts are more intense than those of members who use the Basic Beachscape program. The Advanced Beachscape effort will include mapping coastal physical processes (analogous to the U.S. Army Corps of Engineers LEO study) and wildlife sightings. This information will also be included in the database and entered into the GIS. Naturally, Beach Stewards will participate in a more intensive training session.

Surfrider publishes its findings in the State of the Beach report, which summarizes the mapping information gathered by Beachscape volunteers. The information collected for this report, along with the initial data collected by Beachscape volunteers, provides a baseline snapshot of the current state of America's beaches. This information is not only useful in itself, but it will allow Surfrider to track trends in important coastal variables such as the amount of beach access and the extent of coastal armoring.

Surfrider Foundation's program has already affected conservation successes, and the implementation of Beachscape is going well. Selected as the first chapter to pilot the prototype, the Ventura County chapter has successfully mapped the entire 16 miles of coast in its county. The chapter has held several volunteer

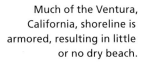

Much of the Ventura, California, shoreline is armored, resulting in little or no dry beach.

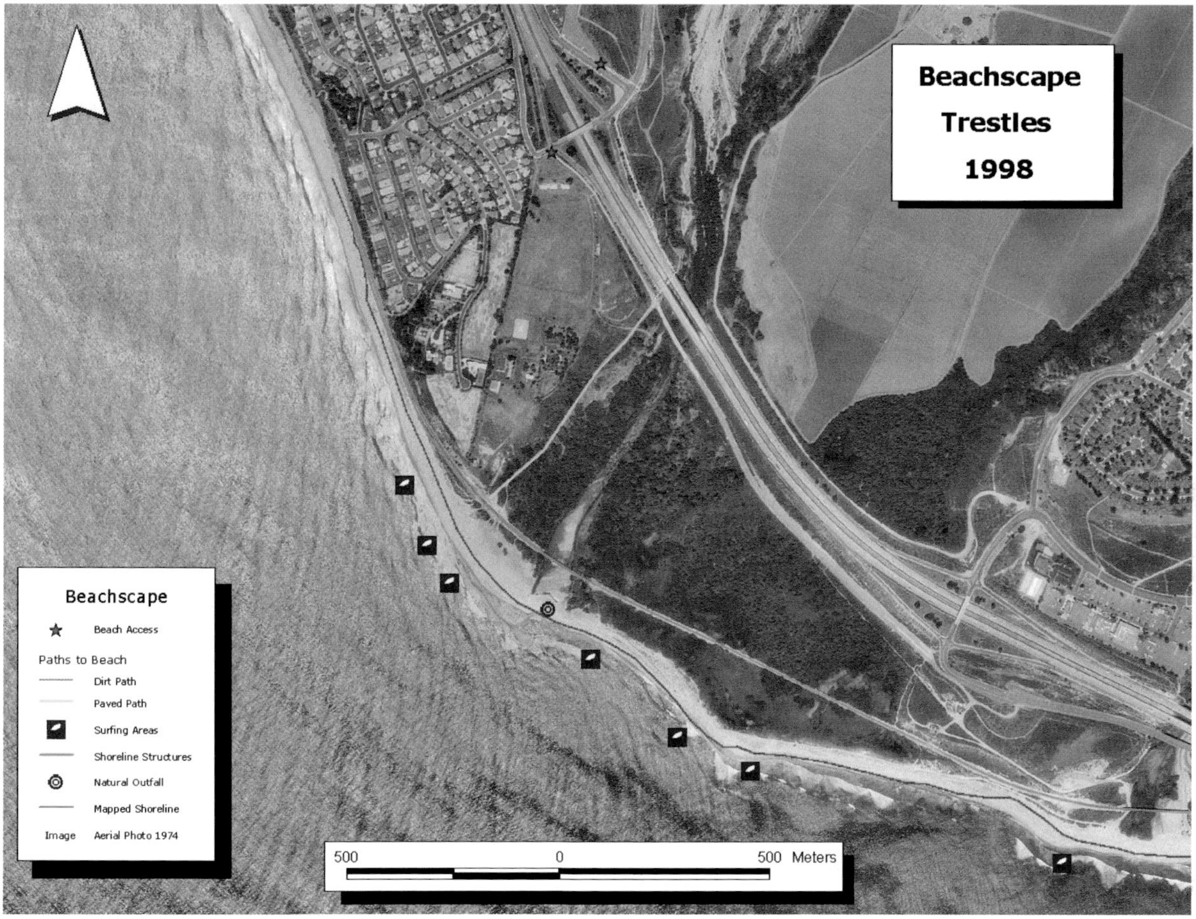

Surf resources are documented along with access to the beach. This view is of Trestles Beach, a popular surfing spot at San Onofre in Southern California.

beach-mapping events to educate the public in Ventura about the importance of monitoring beaches and also about some of the negative consequences of poor coastal management. During these sessions, participants collected a valuable baseline of beach information for the county. More than 90 storm drains, a prime source of ocean pollution, were identified.

The Boston chapter successfully mapped beach access points (or the lack thereof) in its state. The chapter used these beach access maps in a hearing to fight for more public access to Massachusetts coastline.

In the midst of one of the longest beach closures in the city's history, the city of Huntington Beach and Orange County officials were aided by a Huntington Beach chapter member who used a Beachscape map of storm drains to alert officials about storm drains. The Long Beach chapter used a map of city storm drains and the results of a Santa Monica Bay epidemiological study to create a poster that warns surfers not to surf or swim near storm drains.

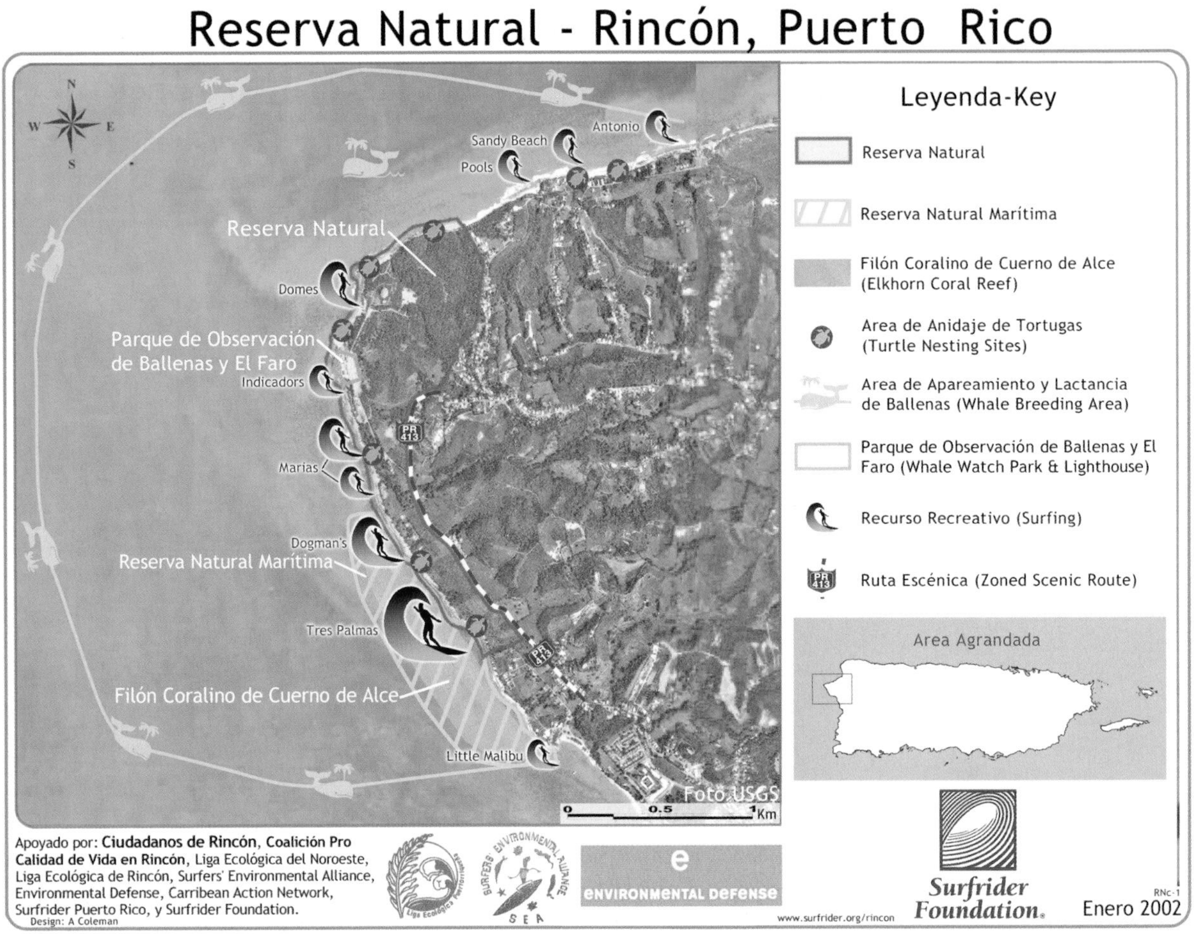

The Surfrider Foundation is active in environmental protection efforts throughout the world, including Puerto Rico.

Having defeated a potentially damaging bike path and seawall proposal in San Clemente, California, the San Clemente chapter is now working with the city to take a proactive look at coastal and erosion management in San Clemente. Ninety percent of San Clemente's coastal area has already been mapped using Beachscape.

Through the use of its effective volunteer programs, the Surfrider Foundation has improved the availability of data along the coast, strengthening the potential for effective coastal management. The foundation continues its progressive mapping program efforts. Surfrider and its partnership organizations are creating data standards that promote further sharing of spatial data.

THE AUSTRALIAN NATIONAL UNIVERSITY

Marine Geography and the Benthic Habitat

Domains of the Australian Ocean Territory

Robert V. Burne and Christian A. Parvey
Department of Geology
The Australian National University
Canberra, Australia

We report the development of a GIS-based project to assist with the rapid identification of a representative system of protected areas for the benthic habitat domains of the Australian Ocean Territory (AOT). Rather than adopt a single bioregionalization approach, we chose to consider the habitat domains of the AOT by developing six geographic data sets indicative of the environmental regimes determining biogeographic variation. The geographic data sets are: bathymetry; submarine physiography; marine substrates; oceanographic conditions at the seafloor; exposure frequency of shelf depths over the last glacial cycle; and the rise of sea level since the last glacial maximum. These geographic data sets are not direct indicators of environmental regimes but are surrogates based on interpretations of incomplete data sets gathered for other purposes. They provide a satisfactory tool for the rapid, small-scale / large-area assessment of the variety of benthic habitat in the AOT. The geographic data sets can be combined and analyzed using GIS technology to provide a variety of insights into the linkages between biogeographic variation and benthic habitat distribution essential for the selection of a representative system of marine protected areas.

Introduction

The adoption of the United Nations Convention on the Law of the Sea (1982) resulted in Australia acquiring rights and responsibilities over some 16 million square kilometers of ocean, known as the Australian Ocean Territory (AOT) (Zann 1996). The major part of this area is under the jurisdiction of the Commonwealth Government, which has consequently been required to rapidly develop an integrated planning and management framework for this vast ocean area. To assist with this process, the Commonwealth Department of the Environment, Sport and Territories commissioned a project to enable the rapid appraisal of the variety of benthic marine habitats and the identification of a representative system of protected areas within the AOT.

Technical background

The platform selected to develop the Marine Benthic Habitats GIS was ESRI's suite of ARC/INFO® and ArcView software. All data was assembled as either ARC/INFO coverages or ARC/INFO grids. Data sets were organized into a directory structure that permitted the rapid development of an ArcView project for visualization, data exploration, and mapping.

Data was geographically rectified, quality controlled and verified, and projected onto an Albers Equal Area projection, with standard parallels of –18.0 and –32.0 with a central meridian of 135.0. This projection was seen as the most suitable for the large geographic area covered, to sustain area calculations, and to enable slope and depth analyses to be performed. However, given the resolution of data capture and mapping, little if any data can be regarded as having validity at scales other than regional. Most are only suitable for mapping at scales of 1:500,000 and smaller and, as such, little importance was paid to geodetic datums and spheroids. Most coverages use the ARC/INFO default spheroid of Clark 1866. Although processing was carried out on a Sun™ UNIX® platform, an 8.3 file-naming convention was adhered to so that the GIS could be exported to any desktop platform.

Methodology and construction of geographic data sets

In onshore areas, habitat definition has conventionally been based on regionalizations derived from land-systems classifications (Thackway and Cresswell et al. 1995). Similar land-system classification methodology has also been used for biogeographic regionalization of the Australian marine environment (Thackway and Cresswell 1998).

Nix (1982) proposed mapping the environmental determinants that modulated plant response as an alternative to land-system mapping. For onshore areas the domains of plants could be defined in terms of thermal, light, nutrient, and water regimes. This approach provides a powerful tool for assessing the representativeness of places for conservation reservation (Mackey et al. 1988).

There are difficulties in directly translating this environmental regime/habitat domain approach to the submarine environment due to the more complex nature of the subaqueous realm. Light is more variable, and absent altogether beneath the photic zone; water is ubiquitous, but properties such as salinity and oxygen saturation gain greater relevance. Microbial and animal life attain greater relative importance at depth, but there are obvious linkages between deep benthic life and primary production near the surface of the overlying water column. The mapping of environmental regimes in the marine realm is also limited by the sparse coverage of reliable environmental information for the seafloor.

In an attempt to overcome these limitations, it was decided to base our investigation of marine benthic habitat variability on the construction of geographic data sets that could be used as surrogate representations of the spatial distribution of environmental regimes at the seafloor. The geographic data sets constructed were: bathymetry; submarine physiography; marine substrates; oceanographic conditions at the seafloor; exposure frequency of shelf depths over the last glacial cycle; and the rise of sea level since the last glacial maximum. The limited availability of basic data from the Australian marine environment compelled some of these geographic data sets to be derived using surrogate data sets, sometimes incomplete and, on occasion, needing extensive interpolation and interpretation. We outline below the methods used to derive the geographic data sets.

Bathymetry

Perhaps the most fundamental characteristic of marine benthic habitat relates to the topography of the seafloor. As such, one of the most vital requirements for this study was a reliable, detailed, bathymetric data set. Yet, when this study was undertaken (July–December 1996), no such data set had been compiled for the AOT. While the General Bathymetric Chart of the Oceans (GEBCO) and the National Oceanographic and Atmospheric Administration (NOAA) provide worldwide bathymetric coverage, this data is of little use on its own in a detailed study. However, it is very useful in augmenting more detailed but spatially limited bathymetric data sets within a GIS environment. At the time of this study, only two sufficiently detailed bathymetric data sets were available for the AOT. For the sake of simplicity, the two data sets will be referred to after the researchers who compiled them: the Buchanan bathymetry (C. Buchanan, Australian Geological Survey Organisation, AGSO), and the Hamilton bathymetry (N. Hamilton, Commonwealth Scientific & Industrial Research Organisation, CSIRO). Both were compilations of existing data, and, as neither has been (to date) documented, it is difficult to allocate any true reliability values against them. Nevertheless, they were regarded as the best data available at the time.

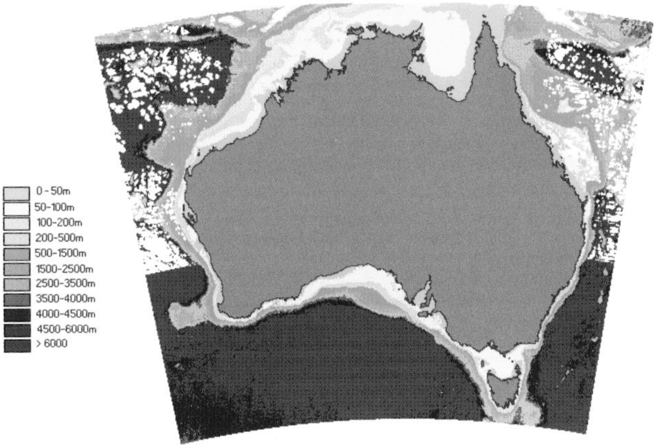

Figure 1: ARC/INFO grid based on the Buchanan bathymetric data set. At the time of this study (1996), this was the most detailed bathymetry for the Australian region.

The Buchanan bathymetry *(figure 1)* was compiled on a 30-second grid and covers an area 9–45° S and 108–157° E . Data was drawn from numerous sources with the best coverage of continental shelf and slope areas. Interpolation away from sounding points or seismic lines was conservative, erring on the side of caution, and, consequently, there are extensive gaps in data coverage. The raw data consisted of approximately fourteen million x,y,z coordinates. These were imported into ARC/INFO and converted to an ARC/INFO grid. The Hamilton bathymetry was also compiled from a variety of sources as a geographic data set for the now-defunct Coastal and Marine Information System (CAMRIS), operated by the CSIRO Division of Wildlife & Ecology (Hamilton and Cocks 1993). Initial processing varied slightly, because the Hamilton bathymetry was not on a specified grid or lattice. Therefore, the x,y,z values needed to be converted to a triangulated irregular network (TIN) to produce a continuous bathymetric surface. The TIN was then converted to a grid with a cell size of 1 kilometer.

	>1:1	1:1–1:50	1:50–1:200	1:200–1:500	1:500–1:700	1:700–1:1,000	<1:1,000
0–50 meters						Photic high-energy shelf	Photic high-energy shelf
50–100 m						Photic low-energy shelf	Photic low-energy shelf
100–200 m						Dark low-energy shelf	Dark low-energy shelf
200–500 m	Upper slope	Upper slope	Upper slope			Deep shelf	Deep shelf
500–1,500 m	Upper slope	Upper slope	Upper slope				
1,500–2,500 m	Middle slope	Middle slope	Middle slope		Continental plateau	Continental plateau	
2,500–3,500 m	Middle slope	Middle slope	Middle slope		Oceanic plateau	Oceanic plateau	
3,500–4,000 m	Lower slope	Lower slope	Lower slope	Rise	Rise		
4,000–4,500 m	Lower slope	Lower slope	Lower slope	Rise	Rise	Abyssal plain	Abyssal plain
4,500–6,000 m	Lower slope	Lower slope	Lower slope			Abyssal plain	Abyssal plain
>6,000 m		Trench	Trench				

Table 1: Interpreted results of slope/depth analysis of the Hamilton bathymetry. Depth classes in meters are tabulated against slope classes expressed as the ratio V:H, where V represents the vertical change and H represents the horizontal distance over which this change occurs. Matrix cells with significant representation in the coverage have been defined as physiographic domains.

Submarine physiography

Australian marine biologists have traditionally considered seafloor habitats of the AOT not in terms of the physiography of the Australian continental margin, but in the context of a model of continental shelf, continental slope, continental rise, and abyssal plain (see, for example, figure 1 in Thackway and Cresswell 1998), derived from the work on the topography of the Atlantic margin of the North American continent (Emery 1966). To overcome this problem, it was decided to produce a submarine physiography geographic data set from a depth/slope frequency analysis of the bathymetric data. In order to perform the slope analysis, a value of 10,000 was added to each cell so that each was a positive number (for some reason, the ARC/INFO grid analysis functions were not reliable when using negative values).

The Hamilton bathymetry covers the whole of the AOT, contains no "holes" in data coverage, and in the deeper offshore areas gave the best picture of general seafloor provinces. An analysis of the frequency distribution of different slope classes at different depth intervals over the entire AOT was undertaken on this data set. A grid was developed classifying depth into 11 classes and slope into six classes (table 1). From this a physiographic model of eight provinces characteristic of the Australian continental margin was derived *(figure 2).*

Marine substrates

Marine sediment data compiled from all available sources by the now defunct Ocean Science Institute (OSI), University of Sydney, represented the best coverage of information on marine substrates in 1996. The OSI database consisted of approximately fifty-three thousand sample points in two primary files. One focused on lithological facies and seafloor biota (a file called CMP), while the other consisted of physical and chemical properties such as grain size and composition (a file called PNT). The data was derived from various sources, including bottom grab samples and sediment cores, as well as other less reliable sampling methods. In order to compile the data into the database, Chris Jenkins of OSI developed an algorithm that parses for key words in verbal bottom sample descriptions (Jenkins 1997). These are in turn registered as numerical values for various geological phenomena into a database, fundamentally converting verbal descriptions to numerical data.

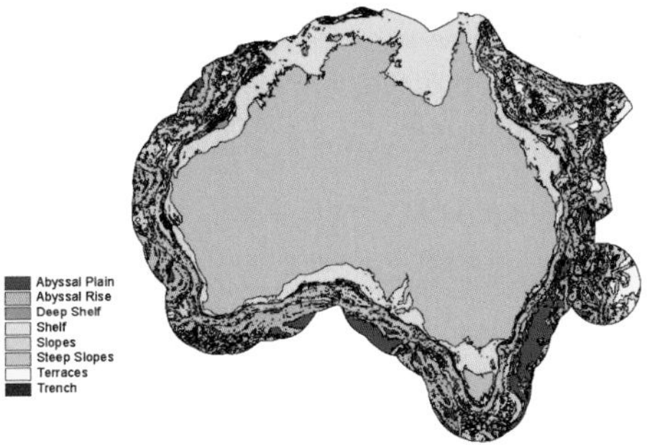

Figure 2: Eight physiographic domains of the Australian Ocean Territory (AOT), based on the Hamilton bathymetry and generalized from the slope/depth analysis summarized in table 1.

Data sets were supplied as ASCII tables. For each sample point, a position was given in geographic coordinates of latitude / longitude decimal degrees. Mapping of the point locations permitted logical checks of positional accuracy to be performed, such as whether a point location fell on the continent (a false location) or at the intersection of Greenwich and the equator (indicating a latitude / longitude location of 0,0, which is clearly a null value). Water depth was also used to check consistency. Any samples failing such tests were rejected.

While the database continued to grow until the closure of OSI in 2001, data used in this chapter reflects the status as of November 1996. While it may appear that Australia's benthic regions have been densely sampled, the distribution of sampling points is very uneven, and large areas of the AOT remain unsampled.

Data on biota in the CMP file yielded some interesting information, but because of the limited number of sampling points it could not be used to characterize the AOT as a whole. For example, coralline material had yet to be recorded in the database from many known coral reef areas. It was decided to use only the PNT file for this study. There are some obvious correlations between sediment characteristics and environmental controls (e.g., finer-grained sediments at depth, carbonate sediments away from sources of terrigenous sediment), but it was decided to undertake a cluster analysis of the PNT data to more rigorously identify sediment classes.

In order to perform the cluster analysis, each sample point needed to have a complete set of attributes, as the algorithm could not cope with "null" values. ARC/INFO was used to select sample points containing measures for water depth, gravel, mud, sand grain size, and percentage of carbonate. This resulted in approximately eleven thousand sample points of useable data.

The cluster analysis was undertaken by K. Malafant using a nonhierarchical method based on the Forgy / Jauncey method using a Gower metric of general similarity measure. This uses a nearest centroid sorting that implicitly minimizes a within-cluster error. The algorithm was developed by K. Malafant (1995).

From preliminary screen plots generated using S-Plus, the optimal number of clusters was deemed to be six or seven. The algorithm was used to generate values for both. The location of samples analyzed and their classifications can be seen in figures 3 and 4. These geographic data sets provide an indication of regions of similar substrate characteristics.

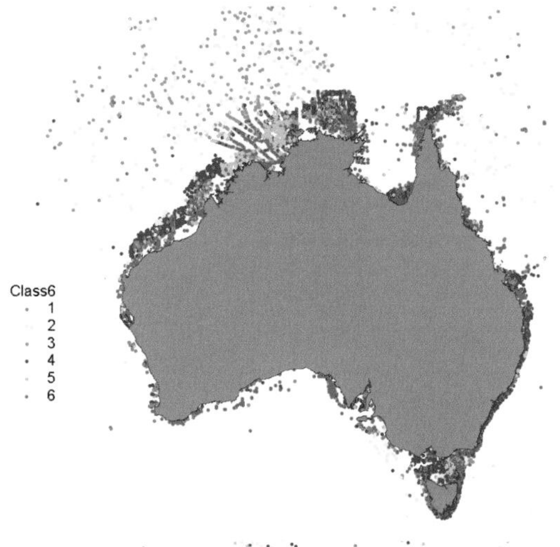

Figure 3: Six classes of substrate domain derived from cluster analysis of selected parameters in the OSI sediment database.

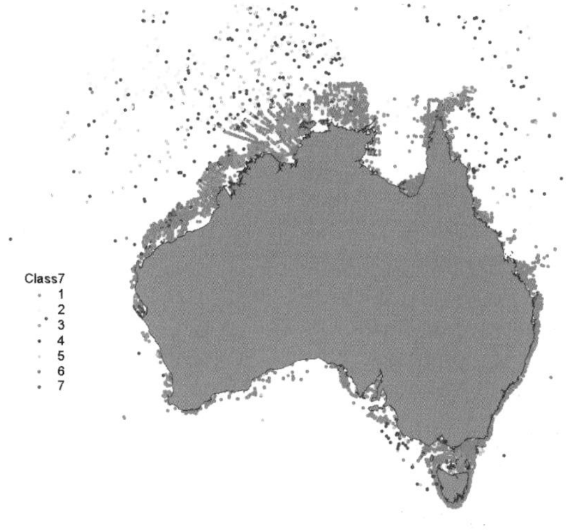

Figure 4: Seven classes of substrate domain derived from cluster analysis of selected parameters in the OSI sediment database.

Oceanographic conditions at the seafloor

The situation regarding oceanographic data from seafloor stations was similar to that encountered with sediment data (Zann 1996). There were several locations that had been extensively researched, but there was no comprehensive, detailed data set covering the AOT at the time of the study. For the purposes of this study, the World Oceanographic Atlas (WOA94) produced by NOAA (1994) was regarded as the best data set available.

WOA94 data is structured around cells of latitude and longitude, providing for each 1° × 1° cell the average annual data for water temperature, oxygen saturation, salinity, phosphate, nitrate, and silicate. Measurements are recorded for up to thirty-three depths to a maximum depth of 5,500 meters. For this study, the greatest depth of measurement at each station was taken to approximate conditions close to the seafloor.

We constructed an ARC/INFO coverage named ANNUALPA, which provides a data value for each of the 2,701 cells covering the Australian region from the equator to 50° S and 100–170° W. When extracting data from the WOA94 database, an algorithm was developed that selected measures for the surface and the deepest value, i.e., the deepest or bottom reading (hereafter referred to as the "bottom"). As part of the coverage ANNUALPA, the value for the surface is given for any property and a corresponding value for the bottom value. The difference between the values for surface and the bottom is also provided.

The variables selected for analysis were temperature, salinity, nitrate, and oxygen saturation. As with the substrate data, cluster analysis was undertaken of all these oceanographic variables by K. Malafant. In this analysis, the initial screen plots indicated that a twenty-group classification would provide maximum differentiation. The resultant geographic data sets of analyzed data are shown in figure 5 for surface waters and in figure 6 for bottom waters.

The important differences between the oceanographic characteristics at the seafloor and those at the sea surface have considerable significance for benthic biogeographic variability. Commonly accepted biogeographic zonations for Australia (Australian Committee 1986) are clearly only applicable to near-surface waters in areas away from the shelf.

Exposure frequency of shelf depths over the last glacial cycle

Quaternary sea-level variations due to the effects of successive growth and decay of land-based ice sheets have been important in determining the biogeography of coastal and shelf benthic habitats. In our original study the late Quaternary sea-level curve of Chappell (1994) was used. For this chapter, a revised composite sea-level curve for the period has been constructed from information published by Chappell et al. (1996), Lambeck and Chappell (2001), and Yokoyama et al. (2000), as well as unpublished information (De Deckker, 2002). The period of inundation of various depths over the past 130,000 years was calculated *(table 2)*. This table was then applied to the Buchanan bathymetry to create a geographic data set showing, for present seabed depths down to −125 meters, frequency of subaerial exposure over the past 130,000 years. Figure 7 is a map based on a specific interrogation of this geographical data set to show the exposed areas for nominated percentages of the last glacial cycle of interest to a particular enquirer.

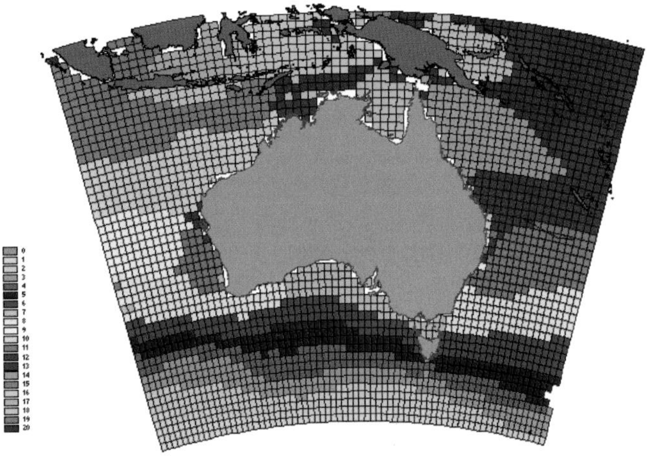

Figure 5: Twenty classes of surface water domain based on cluster analysis of WOA94 oceanographic data.

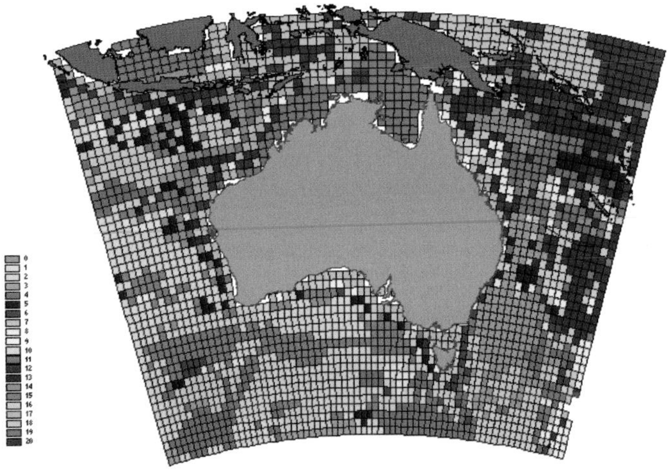

Figure 6: Twenty classes of bottom water domain based on cluster analysis of WOA94 oceanographic data.

Table 2: The percentage of time that various depths of the Australian Shelf were exposed subaerially over the past 130,000 years (the lastest glacial cycle).

Percentage past 130,000 years exposed	Depth below present sea level
80	–13 meters
70	–28 meters
60	–37 meters
50	–44 meters
40	–51 meters
30	–60 meters
20	–68 meters
10	–90 meters
0	–125 meters

Table 3: Australian shorelines relative to present sea level at 5,000-year intervals between 20,000 years and 5,000 years B.P. to show the progress of the transgression since the end of the last glacial maximum.

Time before present	Depth below present sea level
5,000 years	0 meters
10,000 years	–30 meters
15,000 years	–80 meters
20,000 years	–125 meters

The rise of sea level since the last glacial maximum

The rise in sea level from a low of possibly as much as –125 meters at the end of the last glacial maximum 20,000 years B.P. to a level at or above present sea level 5,000 years B.P. (Chappell et al. 1996; Lambeck and Chappell 2001; Yokoyama et al. 2000; De Deckker 2002) needs to be taken into account when considering the present biogeography and evolution of habitat domains in shallow environments. A revised sea-level curve mentioned in figure 5 was used to generate a table documenting the depth below present sea level of Australia's shorelines over the past 20,000 years *(table 3)*. This table was then applied to the Buchanan bathymetry to create a geographic data set showing the progress of the transgression from the last glacial maximum, 20,000 years B.P. Figure 8 shows the decreasing areas exposed above shorelines at successive dates selected to coincide with the timing of an event of biological or anthropological interest onshore.

Note that these data sets are limited by the facts that neither this geographical data set, nor the one generated to show late Quaternary sea-level change, makes any allowance for isostatic response or the effects of contemporary sedimentation.

Discussion

Our study provides an illustration of the suitability of GIS as a tool for small-scale/large-area environmental applications in marine geography. Our approach to the task of rapidly appraising the variety of marine benthic habitats in the AOT differs radically from that conventionally adopted. We chose not to provide a single classification of data into discrete closed polygons, each representing one of a classified group of "habitat zones" (cf. Thackway and Cresswell 1998). Such a classification is based on a single interpretative model of habitat data and can yield very little in the way of fresh insights. In many situations, it proves difficult to connect the classification in any meaningful way to the specific decision-making processes of environmental management. For example, such a classification does little to assist in answering questions such as the delineation of specific environmental limitations revealed by the spatial distribution of a particular species, or the identification of areas of comparable habitat domain for comparative consideration for reserve selection.

Instead, we chose to develop six geographic data sets that would each provide surrogate information on the environmental regimes that form the basis for identifying the range and variability of Australia's benthic marine habitats. The obvious limitation of our approach is that the six geographic data sets generated are themselves interpretations of incomplete data sets. Not only that, but each data set was collected for a specific reason other than that for which it is used in this study. Despite this, the final GIS product has the potential to be used for far greater analysis than can be summarized in the scope of this chapter. By taking these geographic data sets through the integration process into uniform geographic space, under the umbrella of a GIS, a richer sense of classification is possible, with the potential for generating new information products and insights designed to meet the specific needs of the investigator.

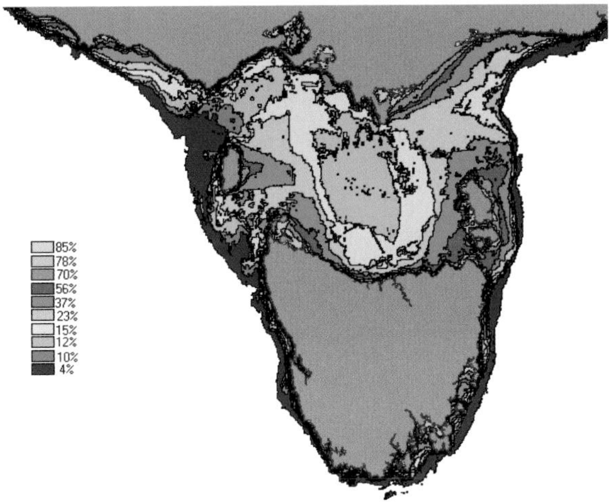

Figure 7: Detail of the areas bordering Tasmania and adjacent parts of Victoria from the geographic data set showing areas exposed subaerially for selected percentages of the past 130,000 years (the latest glacial cycle). The actual percentages selected reflect the interests of a particular investigator.

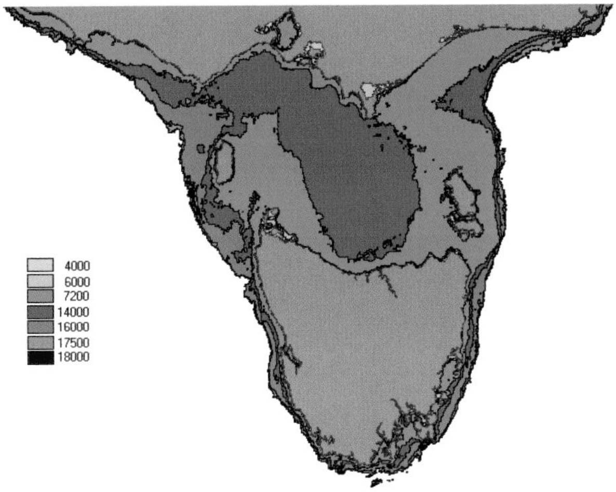

Figure 8: Detail of the areas bordering Tasmania and adjacent parts of Victoria from the geographic data set delineating area of exposure above successive shorelines during the transgression that followed the last glacial maximum (20,000 years B.P.). The shorelines selected correspond to the timing (in calendar years B.P.) of onshore anthropological and biological events of interest to a particular enquirer.

Acknowledgments

The collaboration of C. Buchanan, N. Hamilton, C. Jenkins, K. Malafant, and I. Musto at various stages of this project is gratefully acknowledged. This study was initiated by the Marine and Coastal Geoscience Project, Australian Geological Survey Organisation, and completed at the Geology Department, Australian National University. The manuscript benefited from comments of P. De Deckker and the patient advice of Joe Breman and David Boyles of ESRI.

References

Australian Committee. 1986. Australia's marine and estuarine areas: A policy for protection. International Union for the Conservation of Nature and Natural Resources, Occasional Paper No. 1.

Chappell, J. 1994. Upper Quaternary sea levels, coral terraces, oxygen isotopes, and deep sea temperatures. *Journal of Geography* 103(7):828–40.

Chappell, J., A. Omura, T. M. Esat, M. McCulloch, J. Pandolfi, Y. Ota, and B. Pillans. 1996. Reconciliation of late Quaternary sea levels derived from coral terraces at Huon Peninsula with deep sea oxygen isotope records. *Earth and Planetary Science Letters* 144:227–36.

DeDeckker, P. 2002. Personal notes, January.

Emery, K. O. 1966. Atlantic continental shelf and the slope of the United States. U.S. Geological Survey, Professional Paper 529–A, 1–23.

Hamilton, N. T. M., and K. D. Cocks. 1993. A small scale geographic analysis system for maritime Australia. CSIRO Division of Wildlife and Ecology Working Document 93/9.

Jenkins, C. 1997. Mapping Australia's sea floor. Third Congress of GISs, (AGSO), Canberra.

Lambeck, K., and J. Chappell. 2001. Sea level change through the last glacial cycle. *Science* 292:679–86.

Malafant, K. 1995. *Murray Darling Basin ecosystems analysis, development and application of regionalisations.* National Workshop on Environmental Regionalisation (NRIC) Report, Canberra.

Mackey, B. G., H. A. Nix, M. F. Hutchinson, J. P. McMahon, and M. P. Fleming. 1988. Assessing representativeness of places for conservation reservation and heritage listing. *Environmental Management* 12(4):501–14.

Nix., H. A. 1982. Environmental determinants of biogeography and evolution in Terrra Australis. In *Evolution of the Flora and Fauna of Arid Australia,* edited by W. R. Barker and P. M. J. Greenslade, 47–66. Adelaide, Australia: Peacock Publications.

NOAA. 1994. *World ocean atlas 1994.* U.S. Department of Commerce, Washington, D.C., four volumes and compact disc.

Thackway, R., and I. D. Cresswell, eds. 1995. *An interim biogeographic regionalisation for Australia, version 4.* Canberra, Australia: Australian Nature Conservation Agency.

Thackway, R., and I. D. Cresswell, eds. 1998. *Interim marine and coastal regionalisation for Australia, version 3.3.* Canberra, Australia: Environment Australia, Department Australia.

Yokoyama, Y., K. Lambeck, P. De Deckker, P. Johnston, and L. K. Fifield. 2000. Timing of last glacial maximum from observed sea-level minima. *Nature* 406:713–16.

Zann, L. P., compiler. 1996. The state of the marine environment report for Australia, technical summary. Canberra, Australia: Department of the Environment, Sport and Territories.

The Baja California to Bering Sea Priority Areas Mapping Initiative and the Role of GIS in Protecting Places in the Sea

MCBI
MARINE CONSERVATION BIOLOGY Institute

Lance Morgan and Peter Etnoyer
Marine Conservation Biology Institute
Redmond, Washington

In many ways, conservation in the sea is no different than conservation on land; protecting places (and thereby the genetic, species, and ecosystem diversity associated with them) is a more comprehensive, cost-effective, and politically viable strategy than imposing separate regulatory regimes on individual species. Growing recognition of this reality has led to a fresh paradigm for conserving marine biodiversity: marine protected areas (MPAs). Interest in this new approach has increased dramatically in recent years as the conventional approach of "command-and-control" regulation has failed to stem the tide of biodiversity loss and fisheries collapse.

Marine Conservation Biology Institute (MCBI), a nonprofit organization based in Redmond, Washington, with a public policy office in Washington, D.C., is dedicated to advancing the science of marine conservation biology and promoting cooperation to protect and restore earth's biological integrity. The organization has played a leading role in fostering the MPA paradigm. In January 2000, MCBI held a scientific workshop that culminated in President Clinton's Executive Order to set up the U.S. government framework to establish a national system of MPAs. In accord with its mission to promote the integration of sound conservation science into marine policy, MCBI is currently focusing on science-based projects to assist efforts to designate marine protected areas.

Without a map, do we know where we're going?

Successful place-based strategies require identifying conservation targets, so the first logical step is to produce maps of the most important places to protect. Along the Pacific coast of North America, where a movement has emerged to establish a network of MPAs extending from the waters of Baja California to the Bering Sea, no such comprehensive, regionwide map exists. In some areas researchers have collected many types of geological and geophysical data, including seafloor bathymetry, sidescan sonar images, sediment and rock types, active fault zones, and submersible-based observations and measurements. Collectively this information represents one of the most extensive marine geologic databases in the world, one that could be enormously useful in planning and designing MPAs. However, most of this data is not integrated in a geographic information system

(GIS) that could be utilized to characterize, classify, and predict the distribution of geological features and associated biological features.

MCBI has launched an initiative to integrate these data sets and supplement them with data from other less-studied areas to produce a scientifically credible map of delineated priority areas that will make the otherwise vague and abstract underwater places tangible for the stakeholders. Collaborators in this multiyear effort are the North American Commission for Environmental Cooperation (CEC) and the Baja California to Bering Sea Marine Conservation Initiative (B2B). The goal of the project is to generate a user-friendly GIS with interpretations that are comprehensible for lay audiences as well as the scientific community.

The challenges of generating the "B2B Map," as it's been dubbed, are not insignificant. It will span a vast geographic scope with a variety of jurisdictional boundaries, and at present data is disparate and inconsistently available. Fortunately, GIS researchers have developed several analytical proxies that provide the opportunity to substitute known physical variables for unknown biological characteristics. Topographical complexity can substitute for species richness based upon species/area curves (Etnoyer 2001). The density of sea surface temperature fronts can be derived from Advanced Very High Resolution Radiometer (AVHRR) data to predict bluefin tuna distributions in the Gulf of Maine (Schick 2002) and swordfish distributions in the North Atlantic (Podesta 1993). Biggs et al. (2000) showed that 80 percent of sperm whale sightings in the Gulf of Mexico occurred in areas of lower-than-average sea surface height. Thus, integrating coastline complexity, sea surface temperature, and altimetry data holds promise for understanding oceanographic processes and steering management efforts in the Exclusive Economic Zones of the NAFTA countries.

The significance of another key type of marine data was recently demonstrated by the discovery that deepwater rockfish assemblages are remarkably alike (species distributions) in similar bottom habitats from central California to Alaska, extending from latitudes of roughly 36–55 degrees (Yoklavich et al. 2000). Findings such as this underscore the value of bathymetric and seabed classification data in identifying and assessing priority habitats. Because physical data can serve as a proxy to indicate the locations of certain types of bottom communities, it is a highly useful complement to patchy biological data of varying quality. Furthermore, in areas of intense fishing activity or other disturbance to biological communities, extensive biological sampling might not lead to any useful results.

Assembling the pieces and bridging the gaps

The structure of the GIS analysis itself will be guided by available information, and the results of a workshop held in Monterey, California, in May 2001. At this meeting, scientific experts informed MCBI that four basic data layers (bathymetry, sea surface temperature, currents, and primary production) provided sufficient information to evaluate benthic and pelagic habitats. These types of data are widely available for the entire region, although they vary in scale and resolution. Using a hierarchical approach we can identify the lowest common denominator for each data type, and nest data sets with high-resolution information into this coarser, overall framework. Fortunately, several other types of information are also presently available, but these data sets are scattered throughout numerous governmental, nongovernmental, and academic organizations.

The ETOPO2 data set from the National Geophysical Data Center (NGDC) and ArcGIS Spatial Analyst render the global seafloor at better than 4-kilometer resolution, permitting easy visualization of seafloor features like seamounts and submarine canyons. Surface current information is one of the most sought after and difficult types of information to collect *(figure 1)*. Marine science offers many tools to understand these processes, including moored buoys, surface and subsurface drifters, satellite altimetry, and circulation models. The one-sixteenth-degree U.S. Navy Layered Ocean Model (Hurlburt and Thompson 1980; Wallcraft 1991; Metzger and Hurlburt 1996) is a six-layer global reduced-gravity thermodynamic "deepwater" model that uses the 200-meter

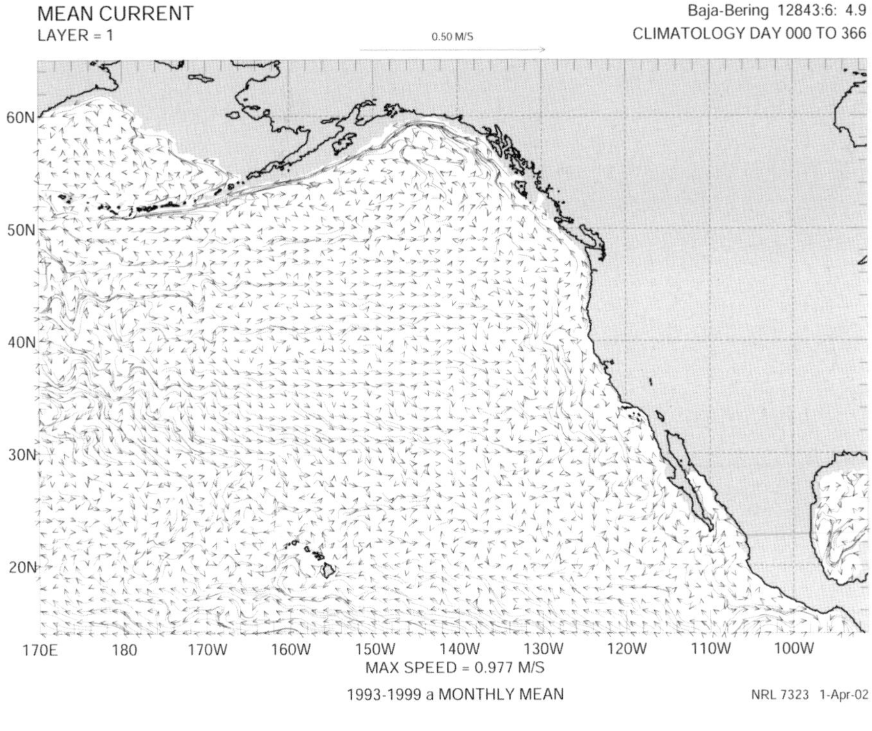

Figure 1: The parameter-driven, one-sixteenth-degree U.S. Navy Layered Ocean Model represents a particular challenge to the B2B mapping effort because it uses rectangular pixels. Vector length indicates predicted monthly current velocity using data averaged over six years.

isobath as the land–sea boundary *(figure 2)*. However, the NLOM uses rectangular pixels, whereas ArcGIS uses only square pixels. Integration of this and other circulation models into the B2B Map will require a unique GIS toolset. Coupling these data sets will allow us to begin to relate aspects of topography to oceanographic processes.

The MCBI-led B2B mapping project will strive to integrate the efforts of a number of other NGOs and government agencies that have begun priority habitat-mapping projects of smaller scope in areas along the west coast of North America. These groups include World Wildlife Fund, Living Oceans Society, The Nature Conservancy, Conservation International, Canadian Parks and Wilderness Society, Channel Islands National Marine Sanctuary (CINMS), and the U.S. Federal Fisheries Management Councils, as well as many others. An appreciation for the value of a single robust map of the region is growing among these groups. This map will not only serve as an excellent informational resource, but it will highlight those special places in the sea that remain anonymous to the public due to the blue veneer of the ocean surface.

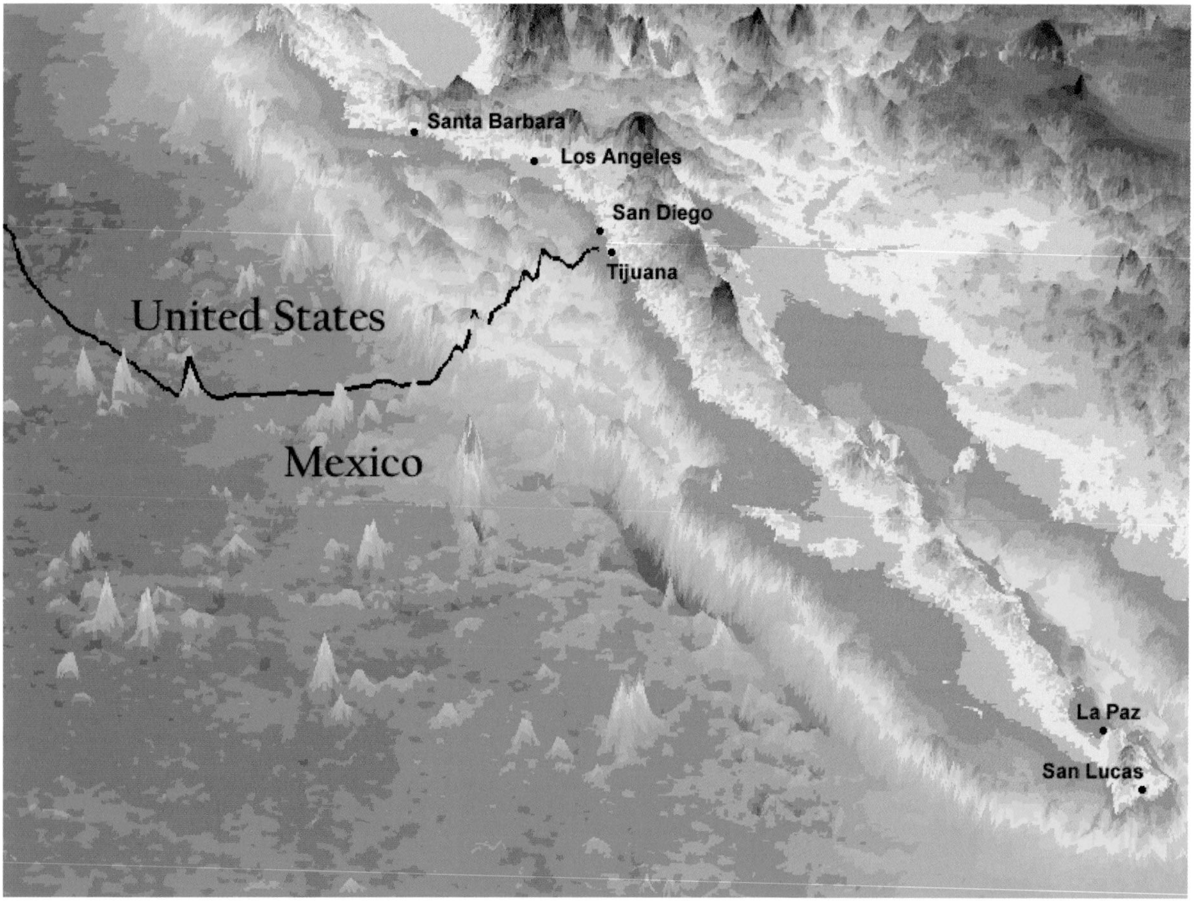

Figure 2: The U.S. Exclusive Economic Zone extends 200 miles off the coast of California. ETOPO2 data from the NGDC reveals the complex bathymetry underlying this maritime border.

To encourage information-sharing opportunities, MCBI is co-hosting a "data potluck" in conjunction with Surfrider Foundation, EcoTrust, and the CEC. The goal of the potluck is to bring together the GIS programs from both regional NGOs and the different local and federal governments in an atmosphere that distributes the workload of data generation and cultivates open relationships among GIS users.

The B2B priority areas team faces several tough questions.

1 What are the appropriate spatial scales of analysis for a continental priority habitat?

2 How do we incorporate previous priority setting efforts for the region?

3 What happens if an area already has many designations (e.g., CINMS)? Does that make another designation more important or less important?

4 How do we deal with the role of various threats in assigning priorities?

5 How do we deal with poorly known regions with little data?

To answer some of these questions, MCBI will supplement physical and biological data sets with a third category, social data. Social data is widely available in many forms, from population censuses and harbor records to regulatory boundaries and levels of activism. We anticipate that this information will prove equally valuable to the final analysis.

Assembling the pieces and bridging the gaps

We believe this broad-based, inclusive process will provide a map that combines scientific credibility and political benefits to the maximum extent possible. It will both generate the best possible product and create a sense of ownership among the participants. If we are correct, the resulting map will serve as a foundation and stimulus for local, national, and international discussions on MPAs, fisheries conservation, coastal zone management, and ocean policy. This process is not intended to designate MPAs or to replace stakeholder processes that do so; rather it aims to provide a robust and objective method for identifying areas of high conservation value.

The planned end product will be a comprehensive, easily accessible, multilayer GIS database, in ERDAS and ArcView formats, that can be distributed on a platform-independent CD–ROM to a diverse scientific and nonscientific community. It will also be distributed in hard copy, and the data sets will be made available via the MCBI and CEC Web sites.

MCBI's efforts to promote the integration of science with ocean policy began as the dream of Elliott Norse, a marine ecologist who founded the organization in 1996 and serves as president. Prior to that, Norse worked in Washington, D.C., on the President's Council for Environmental Quality and edited the seminal book, *Global Marine Biological Diversity.* MCBI holds symposia and scientific workshops on emerging marine conservation issues and works to raise public and policy-maker awareness on them. MCBI maintains an extensive Web site at *www.mcbi.org.*

References

Biggs, D., R. Leben, J. Ortega-Ortiz. 2000. Ship and satellite studies of mesoscale circulation and sperm whale habitats in the northeast Gulf of Mexico during GulfCet II. *Gulf of Mexico Science* 1:15–22.

Etnoyer, P. 2001. Sources, sinks and the spatial distribution of coral reefs in two large marine ecosystems. Master's thesis. Duke University Nicholas School of the Environment. Duke University Marine Laboratory, Beaufort, N.C.

Hurlburt, H. E., and J. D. Thompson. 1980. A numerical study of loop current intrusions and eddy shedding. *Journal of Physical Oceanography* 10:1611–51.

Metzger, E. J., and H. E. Hurlburt. 1996. Coupled dynamics of the South China Sea, the Sulu Sea, and the Pacific Ocean. *Journal of Geophysical Research* 101(C5, May 15):12331–352.

Podesta, G. P., J. A. Browder, and J. J. Hoey. 1993. Exploring the association between swordfish catch and thermal fronts on U.S. longline grounds in the western North Atlantic. *Continental Shelf Research* 13:252–77.

Schick, R. 2002. Using GIS to track right whales and bluefin tuna in the Atlantic Ocean. In *Undersea with GIS,* edited by Dawn Wright, 65–81. Redlands, Calif: ESRI Press.

Wallcraft, A. J. 1991. The Navy layered ocean model users guide. NOARL Rep. 35. Naval Research Laboratory, Stennis Space Center, Miss.

Yoklavich, M. M., H. G. Greene, G. M. Cailliet, D. E. Sullivan, R. N. Lea, and M. S. Love. 2000. Habitat associations of deep-water rockfish in a submarine canyon: An example of a natural refuge. *Fishery Bulletin* 98:625–41.

Development of an Integrated Marine and Coastal GIS for Oil Spill Response in Australia

AMSA
Australian Maritime
Safety Authority

Trevor Gilbert and Tracey Baxter
Australian Maritime Safety Authority
Canberra City, Australia

The marine and coastal environments of Australia are of vital importance to the health and wealth of the nation. The coast and near-shore regions are a rich source of livelihood for coastal townships and indigenous communities. They also support the larger commercial fisheries, shipping, and tourist industries. The key feature of the Australian coasts and marine environments is their massive size and diversity covering every climatic sea zone, including sub-Antarctic, temperate, and tropical ecosystems. In recent years many world heritage, marine parks, and marine protected areas have come under the umbrella of the Great Barrier Marine Reef Marine Park—the largest marine park in the world. Up to 80 percent of Australians live on or near the coast (within 50 kilometers of the shore), but up to 70 percent of this coastline is remote and uninhabited, with some unique marine and terrestrial flora and fauna. Shipping is also important to the Australian economy, with approximately 97 percent of all goods exported and imported by sea. Based on cargo and kilometers traveled, Australia is one of the top five shipping nations. It hosts more than twelve thousand ship visits each year around the coast.

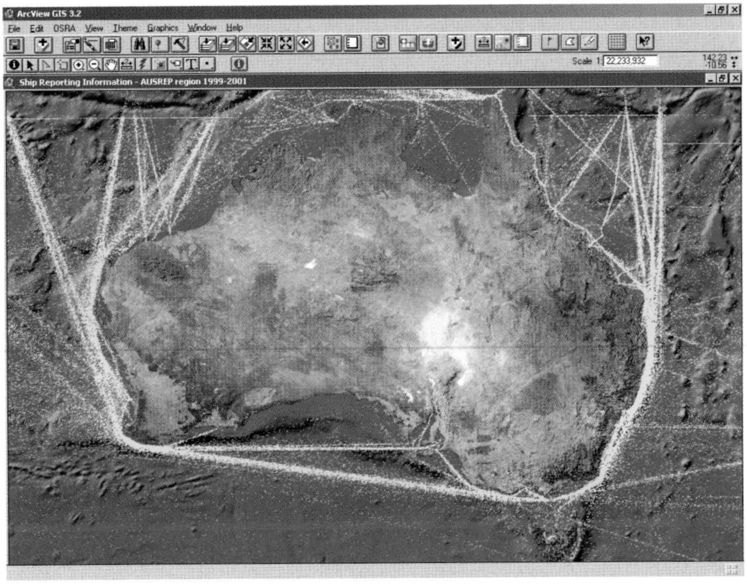

Figure 1: Ship reporting information for vessel traffic in Australian maritime region using ArcView software.

Unfortunately, shipping accidents and illegal discharges of pollutants into the Australian marine environment do occur. More than 1,680 oil discharge sightings and oil spills were reported to the Australian Maritime Safety Authority (AMSA 2001) between mid-1995 and mid-2000. Oil and chemical spills in the marine environment can have widespread impacts and long-term consequences on wildlife, fisheries, coastal and marine habitats, and human health and livelihood, as well as the recreational resources of coastal communities. Since October 1973, Australia has had in place a preplanned national strategy to respond to marine spills from vessels. This strategy is known as the "National Plan to Combat Pollution of the Sea by Oil and Other Noxious and Hazardous Substances," also known as the "National Plan."

It is vital that oil spill response organizations have access to up-to-date information and efficiently organized and user-friendly decision support systems. AMSA, as managing agency for the National Plan, has recently completed an initiative involving the development of a national geographic information system (GIS) to assist oil and chemical pollution response in the marine environment. GIS provides a quick and efficient means of determining environmental, economic, and strategic sensitivities that could be impacted in the event of a pollution incident, and also valuable resource and logistical information. The Australian national GIS system is called the Oil Spill Response Atlas (OSRA).

The major outcome of the OSRA system was the design of an integrated and uniform spill response atlas for Australia in a computerized GIS format. The atlas can be quickly accessed and operated by spill response organizations, planning and cleanup teams, environmental and wildlife agencies, and other emergency organizations.

Applications of GIS to marine pollution response and planning

A GIS provides a consistent view of information throughout spill response operations and provides a means of managing a wide range of data from a variety of sources. It gives managers a fast and effective means of assessing an incident and responding to the specific and changing needs of the situation. It also provides a means of briefing spill response personnel, the media, and the public to provide an up-to-date status and history of response activities.

A GIS can be used for managing information related to prespill contingency planning. It is also useful for monitoring an actual spill and for assessing environmental damage after the spill. A well-developed GIS can be used in marine pollution response to:

- Prepare site/regional contingency plan maps
- Help determine protection plans for shorelines
- Assess habitats affected or likely to be affected by a spill
- Determine species likely to be impacted by marine pollution
- Measure affected shorelines
- Visually represent response strategies and cleanup operations
- Monitor environmental data management
- Calculate area of slicks from field Global Positioning System (GPS) readings or landmarks
- Keep a historic record of equipment locations and deployment

Australia's Oil Spill Response Atlas GIS

The goal in developing the Australian OSRA GIS is to systematically compile all relevant geographic and textual data into a standard national GIS format for the majority of Australia's maritime and coastal environments. Its prime purpose is to assist planners with identifying at-risk resources, and supporting a quick assessment of response options for environmental protection and cleanup.

The national OSRA includes data representing the following elements:

- Biological, environmental, wildlife, and man-made resources for all of Australia
- Geomorphological mapping and shoreline sensitivity to oil spills
- Human-use resource considerations
- Logistical and infrastructure information to support a spill response

The OSRA GIS includes maps, charts, satellite imagery, and point, line, and polygon digital data as well as databases and textual information in a user-friendly, point-and-click format. Data sets that have been acquired and collated for the OSRA GIS include, but are not restricted to, the following:

- Habitats, both coastal and nearshore marine
- High-definition coastlines
- Bathymetry for all Australian waters and high-resolution contours for selected depths
- Nautical charts in scanned, georeferenced raster format
- Scanned topographical charts for all of Australia (1:100,000 scale)
- Marine and national parks and other reserves
- Biological resources and conservation status
- Fisheries and aquaculture
- Coastal and marine wildlife resources
- Recreational resources
- Locations of National Plan equipment stockpiles
- Aerial photography for selected regions
- National Landsat remote-sensing imagery (color 50-meter resolution)
- Oblique photography linked geographically for selected regions
- High-resolution SPOT (Satellite Pour l'Observation de la Terre) imagery for all harbors, ports, and marine parks
- Landmarks and features
- Shoreline access and roads
- Airports, marinas, and boat ramps
- Emergency logistic and other infrastructure information

In OSRA, automation tools have been developed as an extension in ArcView 3.2 to assist OSRA users in the management of data and incident information. It allows quick access to themes for display, analysis of impacts of the spill, and efficient input/output of data and maps. The tools allow users to undertake a range of GIS activities to support oil spill response requirements (e.g., theme manager, locate incident position, import spill trajectory model data sets, create shapes, nearest feature, and incident/event history).

Case study: Use of OSRA GIS in a ship grounding in the Great Barrier Reef Marine Park

In the early morning of November 2, 2000, the Malaysian registered container ship *Bunga Teratai Satu* grounded on Sudbury Reef approximately 15 nautical miles (nm) east of Fitzroy Island and 23 nm from the Port of Cairns within the Great Barrier Reef Marine Park *(figure 2)*.

When it grounded on the reef, the vessel was carrying approximately 1,212 metric tons of fuel oil and 94 metric tons of diesel oil. Although damage was sustained to the vessel during the grounding, the weather conditions remained favorable and no pollutants escaped from the vessel. During the response and the successful salvage of the vessel, the OSRA system was used for contingency planning and operational purposes. The following are examples of the use of the OSRA GIS in this scenario.

OSRA provides efficient storage, retrieval, analysis, and display of environmental and resource information to support a range of uses in oil spill operations and planning. The OSRA Incident wizard starts when the OSRA extension is loaded into ArcView. The wizard either creates a new incident or can display an existing incident and automatically generates an incident number. Next, it will devise a directory structure for all files created during that incident. It will then ask the user to load the OSRA themes in the view and enter the incident location. The user can do this by selecting a position on the GIS view manually inputting geographic coordinates, or by selecting a named location from the gazetteer.

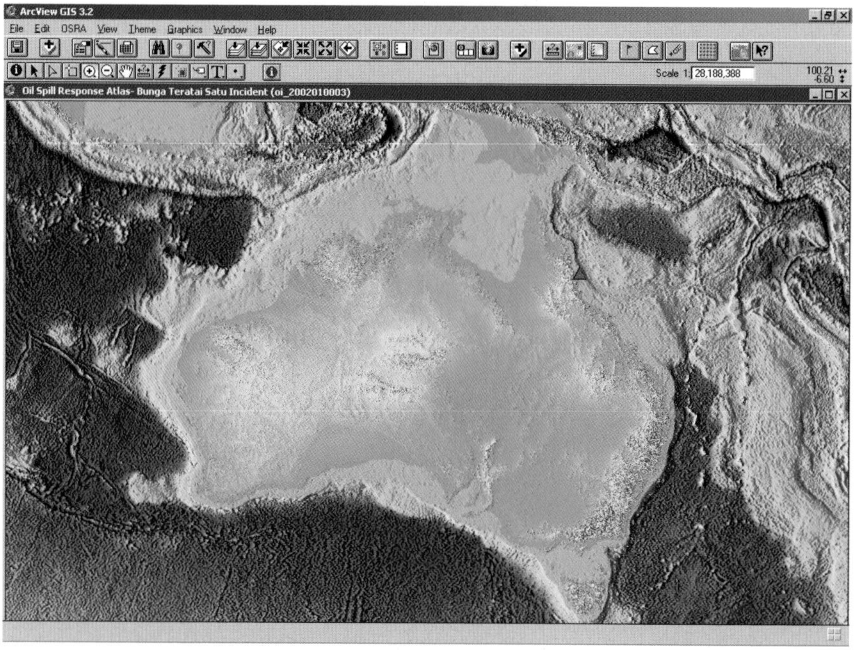

Figure 2: Location of ship grounding in Great Barrier Reef overlaid on bathymetry and topographic data. The orange triangle in the northeast section of the continent identifies the location.

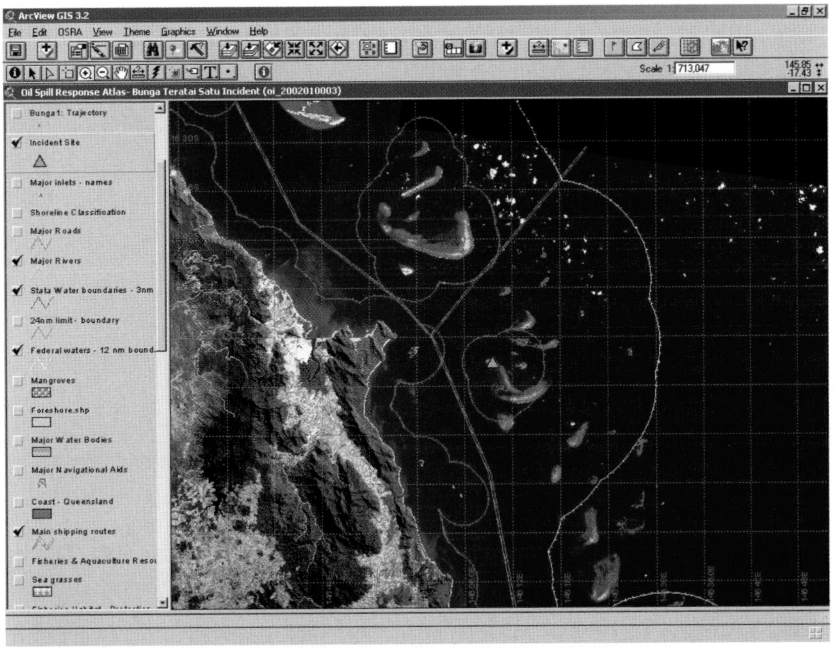

Figure 3: Ship grounding position and major shipping lanes overlaid on OSRA Landsat image.

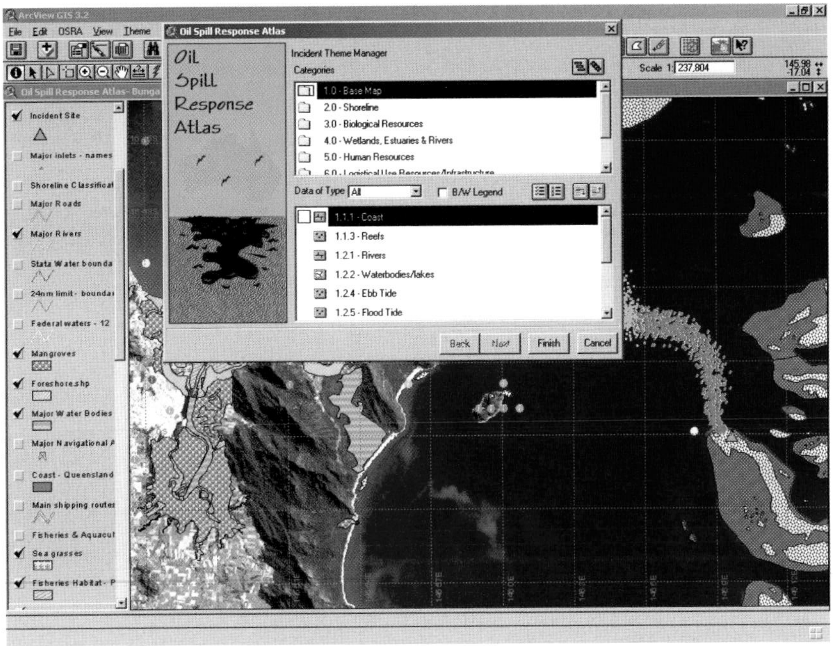

Figure 4: OSRA Theme Manager tool allows the user to select an additional theme.

The theme manager allows the user to load and display only those GIS layers (themes) necessary for supporting decisions in the region of the incident and at whatever scale the user requires *(figure 4)*.

One of the major advantages of the OSRA GIS is the ability to allow the oil spill trajectory model, run by AMSA in Canberra, to be sent as an e-mail attachment directly to remote locations. The user can display the model with the incident information and environmental data on screen *(figure 5)*. The OSRA system also incorporates the complete set of scanned and GIS-ready nautical charts for all of Australia. This includes those nautical charts available for offshore sites and remote Australian islands and territories.

In oil spill response the most commonly asked question is, "Where will the oil go and what resources will it impact?" The Nearest Feature tool allows users to use the attributes and latitude/longitude in the data to select those environmental resources that will be impacted by the oil spill *(figure 6)*.

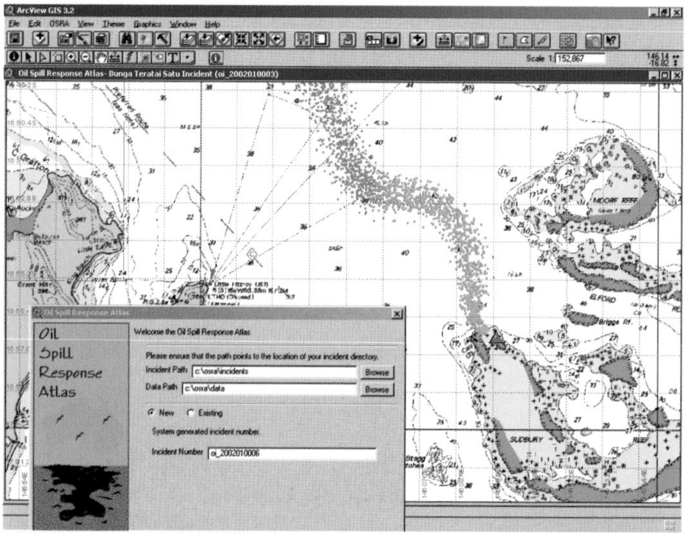

Figure 5: Overlay of spill trajectory model on nautical chart for the incident scene. Ship position is indicated by a red triangle on Sudbury Reef.

Figure 6: The oil spill model is overlaid on a Landsat image with sea grasses and fisheries habitat polygon data. The Nearest Feature tool allows users to select themes and identify the feature nearest to the spill origin or spill trajectory.

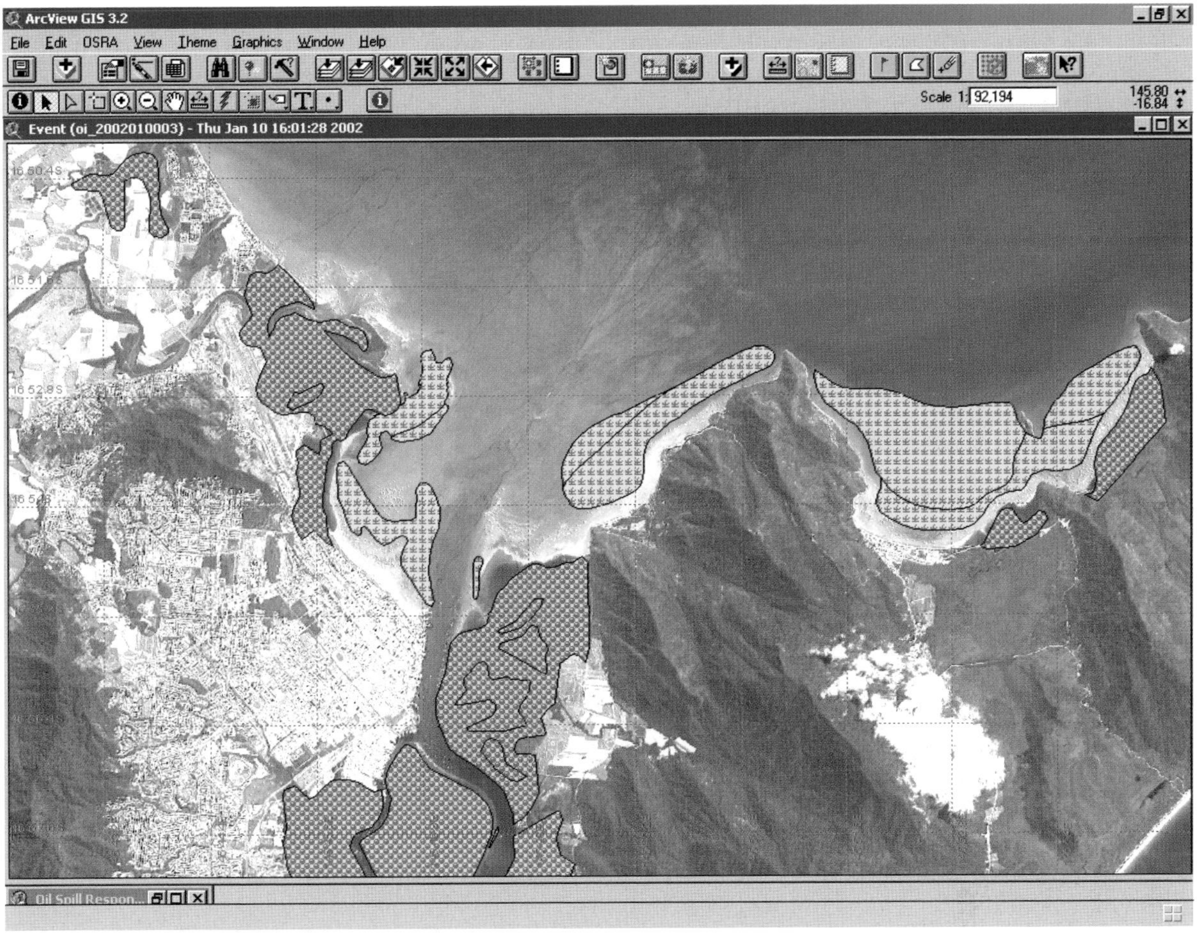

Figure 7: Example of the detail available in the SPOT imagery for the nearby port of Cairns with mangrove and sea-grass beds mapped.

High-resolution SPOT satellite imagery for all major ports, harbors, and environmentally significant areas of the Australian coast has been obtained and loaded onto the OSRA GIS *(figure 7)*. The complete set of national topographical maps (100,000) are also loaded, and are in a georeferenced format.

The toolset also allows users to create points, lines, and polygons for common oil spill response requirements. These include areas designated for dispersant spraying operations, overflight/surveillance of spills at sea, location of boom placement, location of contaminated foreshores, equipment location, and personnel *(figure 8)*.

During the spill, users can also log events into the OSRA GIS for audit purposes and future analysis. Digital images and other photographic information about the spill are also easily logged into the GIS *(figure 9)*.

Figure 8: Image display posits the location of where dispersant could be applied if the vessel had leaked oil. The tool allows users to either upload GPS readings or manually enter coordinates and save these as additional data files for the incident.

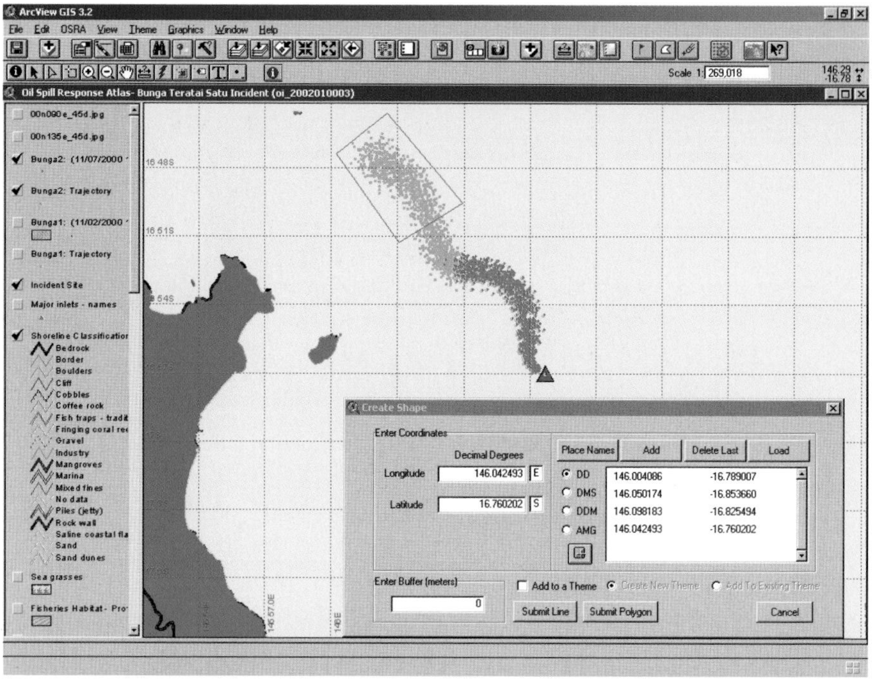

Figure 9: Overflight photograph of the ship grounded on Sudbury Reef hot-linked and georeferenced to the nautical chart.

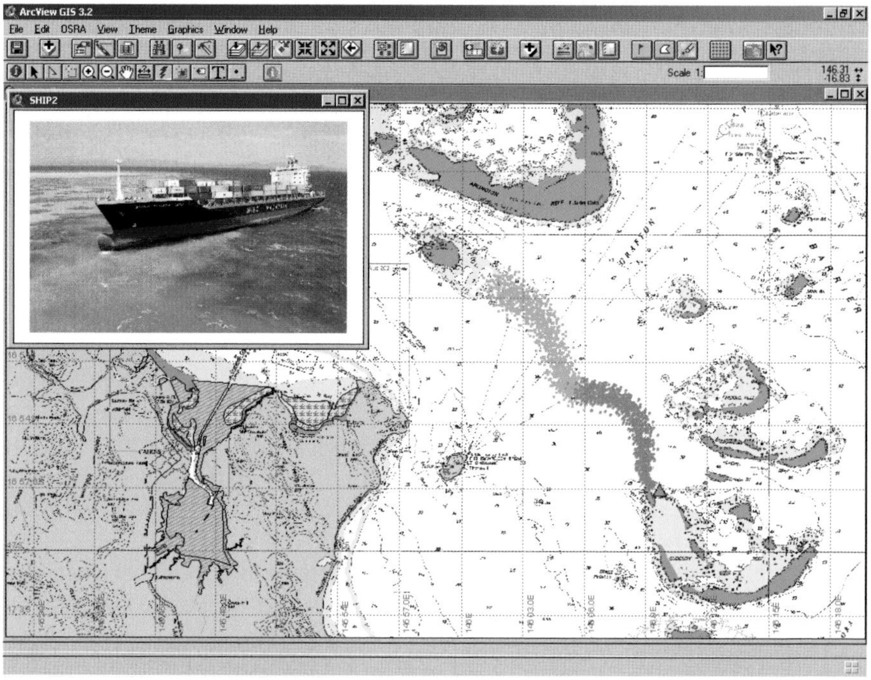

Summary

In maritime emergency response the old adage stands true: Poor information makes bad decisions and good information makes better decisions. It is vital that oil spill response organizations have access to good information and well-organized decision support systems. The ArcView-based OSRA GIS provides a more effective means of accessing a large array of databases, charts, tables, photographs, and text information for the continent's vast coastlines and integrating it all into one easily understood format for responders—a map.

The OSRA initiative now provides Australian state, territory, and commonwealth agencies with vital environmental resource and logistic information in a direct and easily manageable form. GIS supports the efficient and effective response to oil and hazardous chemical spills.

Reference

AMSA. 2001. Annual Report 1999–2000 of the National Plan to Combat Pollution of the Sea by Oil and Other Noxious and Hazardous Substances. ISSN: 1323–7772.

Northeast Pacific

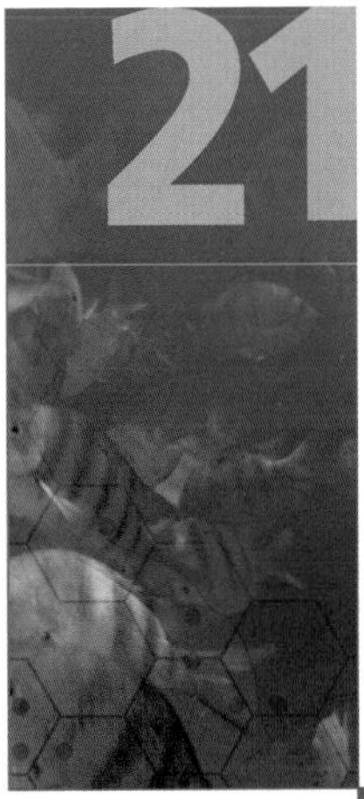

Approaches to Integrating a Marine GIS into the Nature Conservancy's Ecoregional Planning Process

Zach Ferdaña
The Nature Conservancy of Washington
Seattle, Washington

The concept of creating a conservation blueprint begins by taking a wider view, having a vision of the landscape, and translating that vision into a well-thought-out plan for conservation. The Nature Conservancy embraces this vision as the long-term survival of all viable species and community types through the design and conservation of portfolios of sites within ecoregions (Groves et al. 2000).

A portfolio of sites is most often thought of as a set of priority areas in the terrestrial environment that have been analyzed for viability by size, condition, and landscape context. This regionally assembled yet locally detailed vision is just beginning to incorporate the marine environment.

The ecoregional planning process involves identifying conservation targets: those species and communities that best represent the landscape of the ecoregion; those that are imperiled; or both. Once a comprehensive list of targets is established, the task becomes heavily data-driven: collecting spatial data, integrating it into a geographic information system, and running the data through different analytical methods.

How the marine environment fits into traditional ecoregional planning

Methods have been established for selecting and transforming plant, animal, and ecological community data around terrestrial conservation targets and integrating it into a GIS. Aggregated point data and community cell-based data have been used in site-selection algorithms that produce priority conservation sites. But with the growing interest in marine GIS, work is now being conducted to see if traditional ecoregional planning techniques and data transformations can be translated to marine conservation targets and their associated spatial data formats.

Data development of marine conservation targets

There are four general spatial patterns to consider when integrating marine communities and systems data within the ecoregional planning process:

1 Coarse (>100,000 acres) Matrix Communities and Systems

2 Intermediate (10,000 to 100,000 acres) Large Patch Communities and Systems

3 Local (<10,000 acres) Small Patch Communities and Systems

4 Linear Riparian Systems

Ecoregional planning teams have used the linear spatial pattern for riparian areas in the arid portions of the western United States. For the Willamette Valley–Puget Trough–Georgia Basin Ecoregion *(figure 1)* we have added the nearshore system to this last category, which is represented by a spatial mean high-water line attributed with marine riparian information (e.g., overhanging vegetation) and across-shore stratification (supratidal to deep subtidal).

After a thorough investigation of available spatial data for marine conservation targets in this ecoregion, it was concluded that the first credible iteration of the plan was to focus on the nearshore and deep subtidal environments. Deep subtidal was defined as the area down to the –40-meter isobath from mean high water. Comprehensive shoreline data characterizing shoreline morphology, substrate, wave exposure, and biota was available in both British Columbia and Washington State. Called the ShoreZone Inventory, the data sets consist of arcs representing mean high water segmented into homogenous stretches called units. Within each shore unit, the shoreline is further divided into a series of tabular across-shore components from the supratidal to subtidal. With more than 8,069 kilometers (5,014 miles) of shoreline, it was necessary to aggregate the original classification of shoreline substrates into a set of conservation targets that best represented ecological nearshore systems found across the ecoregion. Additional biological modifiers were tagged to each shoreline substrate target in unique combinations of salt marsh, sea grass, and kelp. The result was a list of 39 shoreline habitat targets spanning the Puget Sound and Georgia Basin.

Once we had representative shoreline habitat types, we needed to assemble discrete data on selected marine fish and mammals, seabirds, invertebrate species, and rocky reef habitat. Individual species were evaluated as conservation targets based on the following categories:

- Restricted/endemic: occurs primarily in one ecoregion
- Limited: occurs in the ecoregion and a few other ecoregions
- Widespread: widely distributed in several to many ecoregions
- Disjunct: occurs in ecoregion as a disjunct from the core of its distribution
- Peripheral: more commonly found in other ecoregions

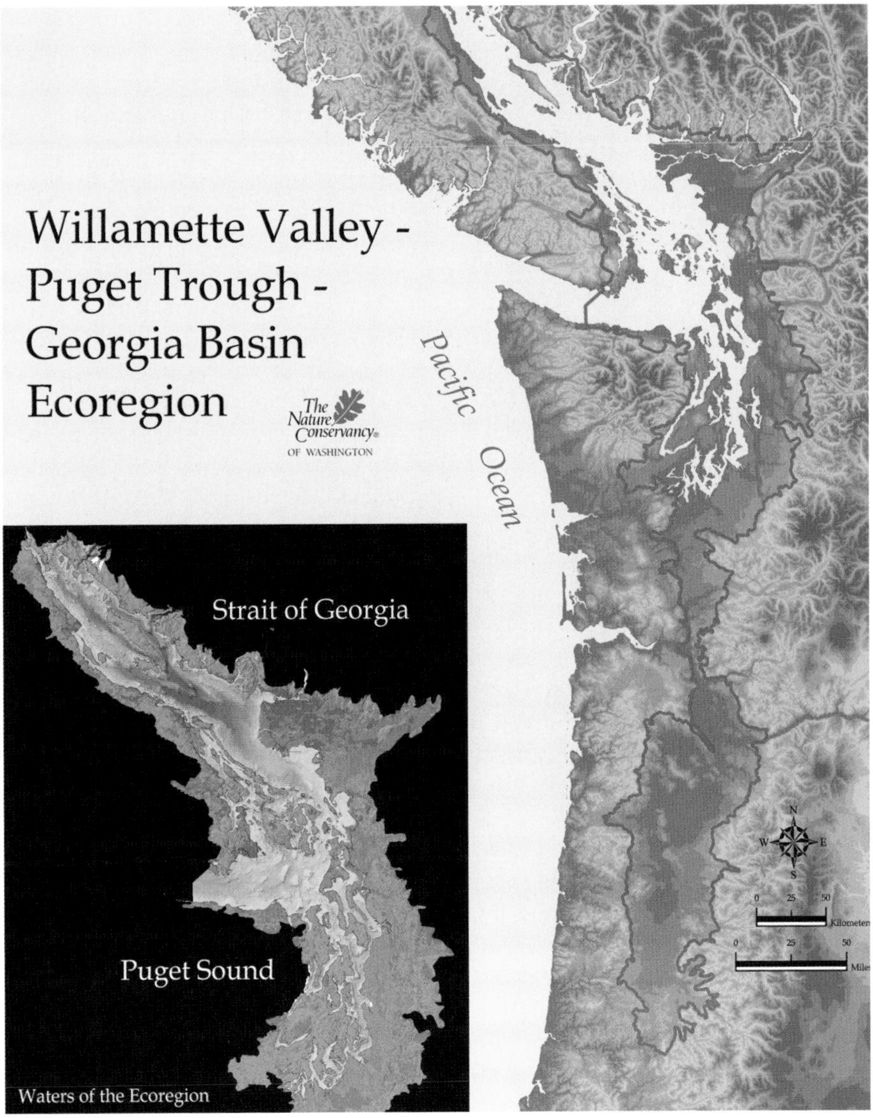

Figure 1: The Willamette Valley (Oregon)–Puget Trough (Washington)–Georgia Basin (British Columbia) Ecoregion contains 8,069 kilometers (or 5,014 miles) of shoreline. This is the first ecoregion where The Nature Conservancy has systematically analyzed the nearshore environment for selecting priority conservation sites.

Generally, more unique, rare, or imperiled species were chosen as targets, as well as those species that were considered keystone species throughout the ecoregion. Although The Nature Conservancy attempts to capture biodiversity within a portfolio of conservation sites, analysis is restricted to available spatial data. Therefore some species, in addition to shoreline habitats, served as surrogates for species not represented.

The next task was to develop methods of spatial data integration between shoreline data sets as well as individual species in order to have consistency throughout Washington and British Columbia.

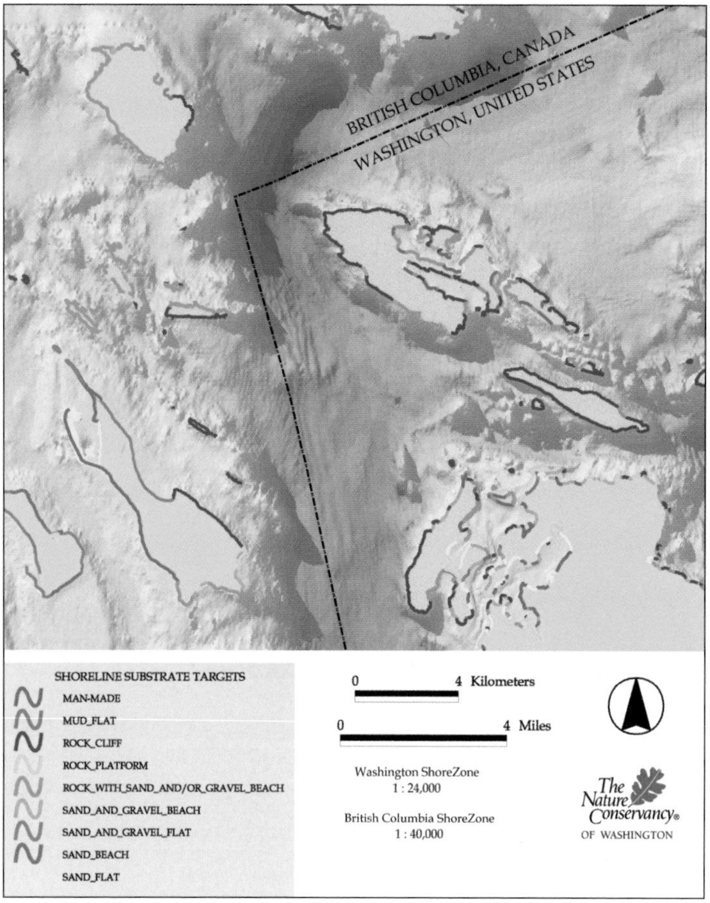

Figure 2: Line simplification was done for the Washington shoreline in order to match the scale in British Columbia when using hexagon sampling units. Original shore units, defined by homogeneous beach substrate, were also used as units of analysis for site selection.

Integrating multiple shoreline data sets for analysis

There are many generalization methods for conducting data integration. For instance, line simplification can be done in a GIS to eliminate nodes in an arc according to a user-defined tolerance. This approach can be used when, for example, you need to accurately measure the length of the shoreline using multiple data sets in multiple scales *(figure 2)*. In using cells or hexagons as sampling units for analysis, we simplified the 1:24,000 Washington shoreline data set to match the 1:40,000 British Columbia shoreline. However, simplification can cause lines to become excessively spiky, leading to serious topological errors especially in complex regions (e.g., estuaries). Therefore, line simplification could be used only with modest scale reduction. Evaluating the results of line simplification was determined by changes in length, angularity, and the number of vertices needed to perform subsequent analysis (Joao 1998).

Another spatial format we used for sampling units involved the original shore segments defined by homogeneous beach substrate. These units varied widely in length, thus providing more natural units for analysis. This characteristic of natural units sets geography and spatial analysis apart from almost every other science (Longley et al. 2001). In this case, line simplification was not done for the Washington shoreline, as nodes were snapped together at 1:40,000, preventing routing along the linear sampling units.

For this ecoregional plan both hexagon and linear sampling units were used to compare the two approaches and evaluate their results in site-selection analysis.

Incorporation of discrete biological and habitat information

The Nature Conservancy, Heritage Data Centers, and the Association of Biological Information (ABI) have developed methods that transform survey data for individual species into Element Occurrences. An Element Occurrence (EO) is an area of land, water, or both in which a species or natural community is, or was, present (ABI 2001). An EO has practical conservation value; it is an area where the element may potentially continue to be present, where it was known to persist in the past, or where it may regularly recur. For species, an EO is generally a local population, but in some cases may be a portion of a population (e.g., long-distance dispersers) or a group of nearby populations (e.g., metapopulation). For natural communities, an EO may represent a stand or patch, or clusters of them. In both cases, an EO represents the fully occupied (or previously occupied) habitat that does or may contribute to the persistence of the species. EOs can also be specific areas for life history functions (e.g., feeding). EOs are typically separated from each other by specific distances across intervening habitat or areas, or by barriers to movement. Minimum separation distances of at least 1 kilometer for species and communities have been established to ensure that EOs are not separated by unreasonably small distances that could lead to fragmented populations being identified.

In lieu of having EO specifications defining EO types for our marine conservation targets, we adopted some of the methods used in creating EOs in the terrestrial environment in order to create concentration areas from survey data. To do this we had to experiment with different generalization techniques that best matched traditional EO concepts.

Generalization for analysis is termed model generalization, and is mainly a filtering process (Joao 1998). This is not used for display, but for data reduction in order to obtain a subset of an original database for analysis. Model generalization aims at minimum average displacement, although some displacement does occur when aggregating multiple survey points into a single point. Filtering survey data into a "marine EO" served to eliminate redundant data elements, speed up subsequent model computations, and adjust the accuracy of different data sets that were processed jointly. In aggregating marine spatial data for an individual species, we found that class aggregation increased attribute accuracy by producing trends. Because a generalized trend over time was more robust than an individual observation, this process was useful for aggregating marine fish, seabirds, and invertebrate data over a number of years, thus increasing the probability of a species to be found at that general location.

We used the field survey units from biologists to summarize the survey data and rank them by either total observations, or in the case of rocky reefs, by relief and complexity *(figures 3–5)*. Putting ranks or conservation values on these aggregated points provided a succinct assessment of estimated viability, or probability of persistence of occurrences of a given element. Their value represents an attempt to provide an assessment of the likelihood that if current conditions prevail, a species occurrence will persist for a defined period of time, typically 20 to 100 years.

After model generalization was performed, we evaluated the effects and placed confidence levels on the data. This assessment of data quality, along with evaluating the importance of a given target, provided a basis for setting appropriate levels of representation for individual habitats and species in the construction of a portfolio of conservation sites.

Figure 3: Video surveys for rockfish and rocky reef habitat were conducted by the Washington Department of Fish and Wildlife between 1993 and 1997. The total number of video drops for rocky reef habitat was 2,315.

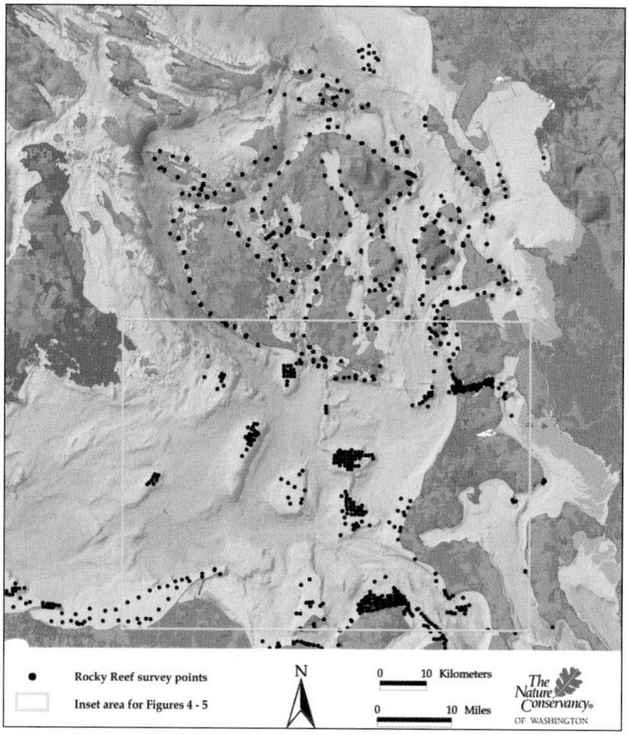

Figure 4: Half-square-mile survey units were used to aggregate the observation data. Relief and complexity of rocky reef habitat was used as the classification for aggregating the data into a single value. The number of aggregated points was 940.

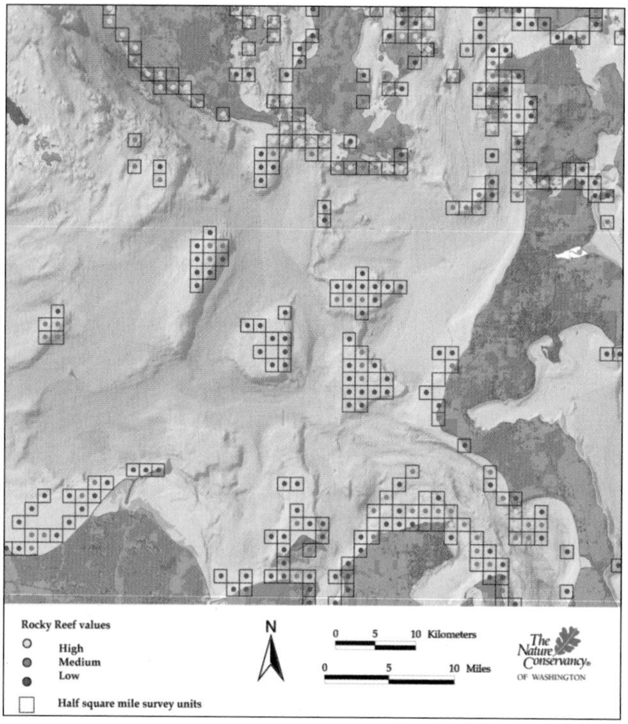

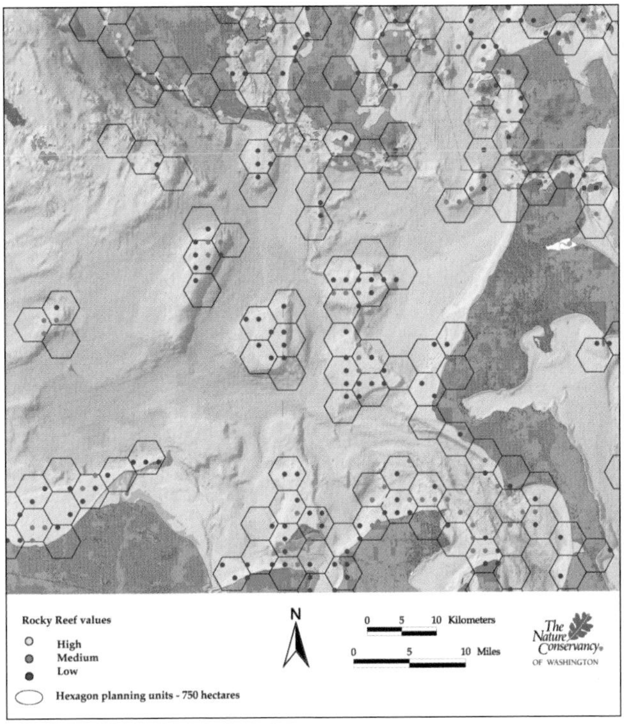

Figure 5: Aggregated points with unique values based on the degree of relief and complexity of rocky reef habitat were summarized into 473 hexagons for analysis. This method served as a standard for aggregating survey point data for marine fish, seabirds, and invertebrates.

The SITES algorithm

The Nature Conservancy funded, through the University of California, Santa Barbara, a project in which Ian Ball and Hugh Possingham wrote a site-selection algorithm for constructing a portfolio of priority conservation sites. This algorithm is called SITES *(www.biogeog.ucsb.edu/projects/tnc/toolbox.html),* and its objective is to identify a set of sites from among a larger set of sampling or planning units within an ecoregion.

Once the spatial data was assembled for our ecoregional plan, levels of representation needed to be applied to each target. This representation, also termed conservation goals, is a percentage of the available spatial data needed to construct a specific conservation portfolio. Adjusting levels of representation, along with other parameters within SITES, changes the results and therefore requires extreme care when evaluating its output. Only through much research and experimentation with the algorithm did a meaningful conservation portfolio result (the final draft of the portfolio will be complete by the summer of 2002).

Comparing two different sampling units

For the first credible iteration of this plan, both hexagons and linear sampling units were used to analyze species and habitat data for site selection. Where all shoreline, discrete marine species, and rocky reef habitat targets were incorporated into the hexagons, only the shoreline and forage fish targets were analyzed using linear sampling units. With hexagons we were able to analyze deeper waters where the linear approach particularly focused on the near shore.

In addition, using hexagons also allowed us to document effects of including both the marine and terrestrial data in the same analysis. The goal here was to see how the algorithm optimized among sites in the nearshore/upland zone.

Preliminary linear analysis produced a portfolio of conservation sites in the near shore more accurately than with hexagons when using only shoreline habitat targets. Sites were more clearly defined because the extents of distinct shore units were the units of analysis. In addition, forage fish spawning grounds were also integrated into the shoreline habitat data because of their spatial proximity in the near shore. However, in integrating biological information from the deep subtidal into the shore units, we decreased the spatial accuracy of that data. Future iterations of the conservation blueprint will advance research methods for increasing attribution of shore units from data in the deep subtidal.

Conclusion

This chapter was intended to describe and illustrate methods for building a marine GIS into ecoregional planning. Having a clear vision for a conservation blueprint will structure the process, from deciphering among the rich marine biota in focusing on key habitats and species, to evaluating appropriate uses of information. This process spatially delineates a set of representative habitat and species assemblages that can be used for a variety of conservation efforts. Subsequent iterations will include more physical parameters such as current, salinity, and temperature in order to study the relationships between an individual conservation site and a reserve network. With this approach, The Nature Conservancy is ensuring that site selection analysis will produce meaningful results and provide stakeholders and decision makers with a reliable plan for conservation.

References

Association for Biodiversity Information (ABI). 2001. Preparation for biotics and implementation of the revised EO methodology, Biotics Team Information Division.

Groves, C., L. Valutis, D. Vosick, B. Neely, K. Wheaton, J. Touval, and B. Runnels. 2000. Designing a geography of hope: A practitioner's handbook to ecoregional conservation planning. *The Nature Conservancy* 1.

Joao, E. M. 1998. *Causes and consequences of map generalisation.* London: Taylor & Francis.

Longley, P. A., M. F. Goodchild, D. J. Maguire, and D. W. Rhind. 2001. *Geographic information systems and science.* Chichester: John Wiley & Sons, Ltd.

22

Marine Mammal and Human Patterns of Use

Inspiring and Informing a Cooperative Stewardship Process

Bob Christensen
Southeast Alaska Wilderness Exploration, Analysis and Discovery (SEAWEAD)
Juneau, Alaska

The first time I camped at Point Adolphus was in 1996. It was the end of September, and I was finishing a three-month kayaking trip in Southeast Alaska, exploring some of the most amazing coastline I had ever seen. Southeast Alaska is one of the few places left on earth where an adventurer can experience the wild on a grand scale. Each cape and cove offers a unique opportunity to expand one's sense of awe and wonder.

Diving whale (near horizon, left) and the sunrise over the Chilkat mountain range.

I arrived at Point Adolphus late in the evening after a tough day's paddle. The reason for this trek to Point Adolphus was that it was near an island where I was to begin a new winter job as caretaker of a remote wilderness cabin. At the time, I had no idea of the area's immense popularity for recreation and tourism, or that it was a site for some of the best whale watching in the world. The day's stormy weather and waning light obscured all such signs.

Once onshore, it quickly became obvious that the beaches around Point Adolphus had hosted many visitors during the summer. Even in the failing light I was quick to discover a path leading to a level spot. During my first night, I was awakened several times by powerful trumpet sounds, carried on the wind. I struggled to guess their source, but could only imagine that they were blasts from a ship's horn sounding through the stormy night. I later discovered that these were the sounds of humpback whales in pursuit of the prey that concentrates in tidal eddies off the nearby point.

The tumultuous weather continued through the week, forcing me to remain at the Point. This haven inspired a sense of wonder as I watched the tides ebb and flow and the symphony of life that played along the shore. The abundance of whales, sea lions, seals, porpoises, and eagles that were a part of my daily life far exceeded my previous experiences in this region. It was then that I began to understand and appreciate the unique qualities of Southeast Alaska.

The conservation challenge

Since that stormy week in 1996, my appreciation for the ecology of the area has only grown stronger. For the past five years, I have made my home on a nearby island but have returned frequently to camp on the Point Adolphus beaches. I am certainly not alone in my feelings for this place. Thousands of visitors come each summer to experience the whales of Point Adolphus. These guests, drawn by reverence for the humpbacks and their habitat, bring the inevitable human encroachments on the environment.

Recreation and tourism pose both threats and opportunities for conserving wildlife and habitat. The rapid growth of the tourist industry combined with the increasing popularity of Point Adolphus point to the human influences of habitat degradation, increased stress on wildlife populations, and social conflicts. Conversely, good management, monitoring of impacts, education, and economic incentives can make recreation and tourism a positive force for conservation.

Much of the concern stems from the lack of guidance that has accompanied the recent increase in recreation and tourism at Point Adolphus. Ironically, for many this lack of guidance is actually part of its appeal. Compared to the highly regulated Glacier Bay National Park nearby, Point Adolphus allows a great deal of freedom for tourism and recreation. During the height of the tourist season, three thousand passenger cruise ships, whale-watching boats, fishing vessels (private and commercial), and other marine traffic are observed daily in the near vicinity of Point Adolphus. In addition to the kayakers, many use the beaches as a convenient base to explore the area and experience the whales, bears, and other wildlife in the marine and terrestrial near-shore zones of Chichagof Island.

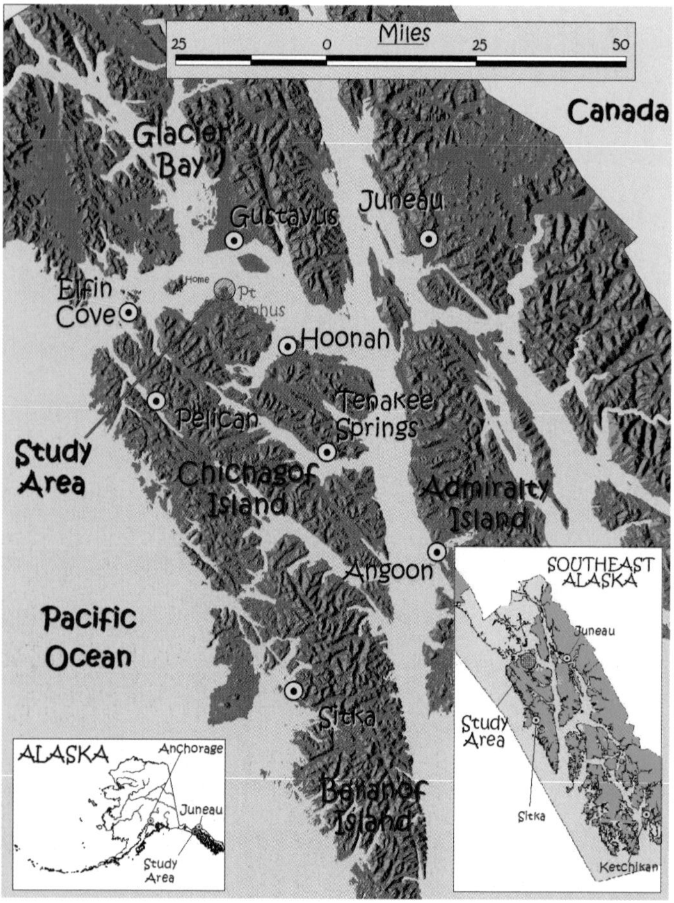

Located at the extreme northern end of Chichagof Island in Southeast Alaska, Point Adolphus extends into the waters of Icy Strait approximately 15 miles west of Hoonah and 10 miles south of Gustavus and Glacier Bay National Park.

Recently, the conservation concerns raised by local citizens, environmentalists, and commercial users escalated. In response, the National Marine Fisheries Service created national regulations for marine mammals. However, the problem of human presence affecting animals still remains. How great is this impact? Even if we could sidestep the quagmire of user conflict, jurisdictional fragmentation, and empty pockets for public education and regulatory enforcement, we would still be without the necessary baseline information to begin an informed exploration of the issue.

A call for cooperative stewardship

Devoted to finding solutions to this ecological problem, a few Southeast Alaskans have worked hard to implement positive changes in the region. In 1999 they formed Southeast Alaska Wilderness Exploration, Analysis and Discovery (SEAWEAD), a nonprofit conservation- and education-based organization. Since its inception, SEAWEAD has worked to facilitate research-based, cooperative stewardship among local citizens, communities, and government agencies of Southeast Alaska.

SEAWEAD's main objectives are to identify conservation issues, collect site-specific scientific and natural history information, and process and present the information in a broadly accessible form.

The purpose of the work performed by SEAWEAD is to bridge gaps between nature and society. We strive to be a focal point for partnerships in conservation. During our work at Point Adolphus, SEAWEAD has benefited from the cooperation and support of Glacier Bay National Park, the University of Alaska, the National Marine Fisheries Service, the U.S. Forest Service, Ecotrust, the Sitka Conservation Society, the Leighty Foundation, the Skaggs Foundation, Alaska Discovery, and the ESRI Conservation Program.

Since 1999, SEAWEAD has been engaged in an intensive study that focuses on the whales of Point Adolphus. The Point Adolphus Humpback Whale Project is centered on understanding the effect of human–whale interactions on the endangered humpback. We employ both stationary observation and mobile survey techniques to collect baseline information about patterns of habitat use at Point Adolphus. The objectives of the project are to: (1) describe the distribution, abundance, and behavior characteristics of humpback whales and humans near Point Adolphus; (2) examine the relationship between humpback whale behavior and human activity; (3) develop scientific reports and education materials to raise awareness of important conservation issues; and (4) provide outreach to local communities, management agencies, and user groups.

The extremes of the human-use spectrum at Point Adolphus: a solo paddler and a cruise ship.

Creating a balanced perspective

In 1999, SEAWEAD began a pilot study focusing on both the stationary observation and mobile survey techniques. The two perspectives blend science and natural history investigation, with the unified purpose of providing helpful, user-friendly information.

Our observation techniques have evolved primarily because of the invaluable support and guidance we have received from Glacier Bay National Park. This technique is structured by the rigors of scientific effort focusing on behavioral responses of humpback whales to human (vessel) presence and activity.

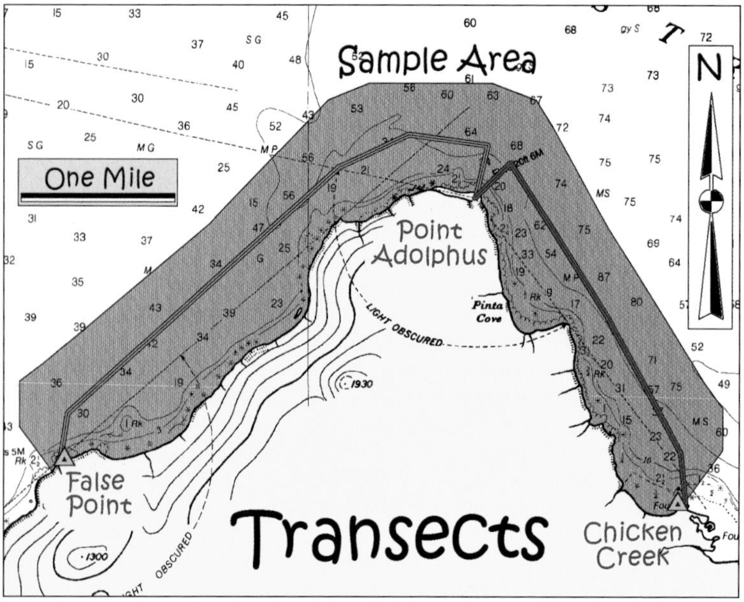

Figure 1: East and west transect locations.

Gathering baseline information

The mobile survey technique of observation provides a broader ecological perspective than the stationary observation. The mobile survey is necessary for describing the distribution and abundance of marine mammal and human activity encountered along approximately 15 kilometers of coastline around Point Adolphus.

From early June to September, we systematically paddle around the coastline of Point Adolphus collecting a variety of information about the patterns of habitat use. We established two 8-kilometer paddling transects *(figure 1)* on each side of Point Adolphus that run within 1.5 kilometers of shore. All marine mammals, human traffic, and campsites were identified and recorded along with time, position, group size, and behavior data. This information provides a baseline description of the distribution and abundance of humpback whales, stellar sea lions, harbor porpoise, sea otters, harbor seals, and human beings using the local habitats.

Results: The heuristic power of naturalist field study and GIS

At the heart of SEAWEAD's outreach efforts is the synthesis of scientific research and naturalist study. Science provides a structured look at critical elements in a system, and the naturalist perspective describes the ecological context for interpreting the scientific data. We believe that when balanced, the two perspectives greatly enhance one another.

The mobile survey technique paints the canvas of what we know about the marine ecology of Point Adolphus with a rich natural history background. We are confident that this background will prove fundamental to understanding and communicating the finer details of humpback whale behavior observed through the stationary observation study.

During the 2000 field season we conducted approximately 60 surveys of transects and recorded 750 data points. Each data point represents a group of whales, seals, sea lions, sea otters, porpoises, or people.

To better understand the output of this effort, imagine that each time we recorded a data point we were able to drop a color-coded buoy to mark the location. The result would look something like figure 2. In order to draw meaningful comparison, I would next run a query in the GIS asking ArcView to show only the points (buoys) where humpback whales and nontransient vessel traffic had been recorded *(figure 3).*

Figure 2: All mobile survey point data.

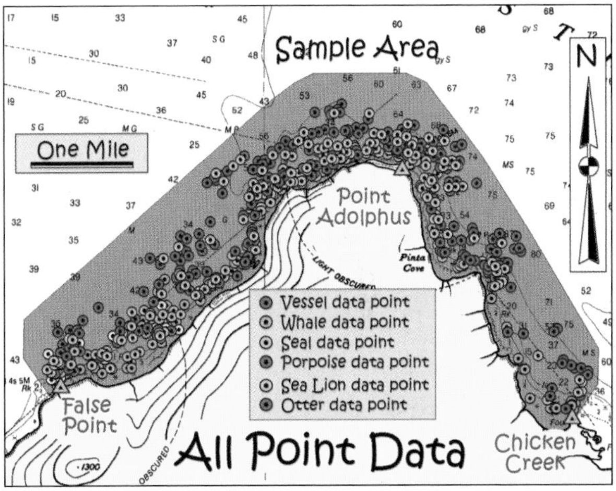

To continue the analysis for determining specie density, visualize that you are floating in among the whale and vessel marker buoys with an imaginary 500-foot pole. If you were to tie up to one of the buoys and reach out with your pole, every like buoy you could touch would be considered an in-buoy. The more in-buoys you gather, the more popular that area is for the species you are counting. We call this the 500-foot pole rule. It is a simplified version of what we ask the GIS program ArcView Spatial Analyst to do for us in processing the mobile survey data. We ask the program to float in every available point (or tie to every in-buoy) within the study area and systematically search within a 500-foot radius for like symbols. The more like symbols that are found, the higher the density calculation (or use level) for the point type.

Mobile survey kayak with "deck-writer" for data collection on-the-go. Data sheets are copies of local chart information with distance delineations and landmark information added in ArcView for sampling standardization.

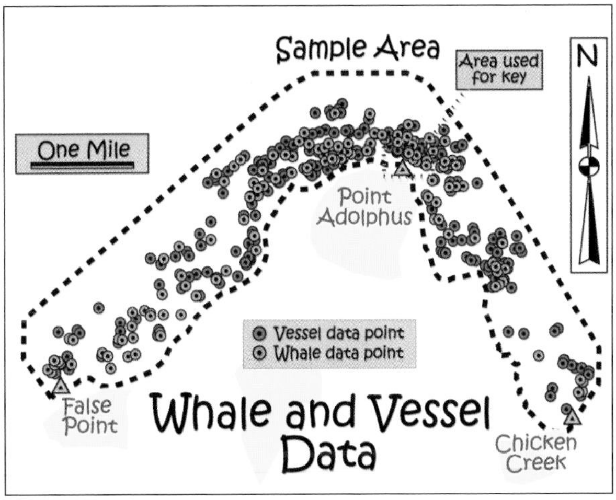

Figure 3: Vessel and whale data points.

Another valuable ArcView Spatial Analyst user-friendly function allows us to represent these density calculations with color shading. To take a closer look at how this shading is done, figure 4 zooms in on the key area indicated in figure 3. Notice how areas with higher densities (use levels) are shaded with bright colors and areas with lower point densities (use levels) are darker. With this map we begin to see a spectrum of habitat-use level for humpback whales encountered on mobile survey.

Figure 4: Densities shown with shading.

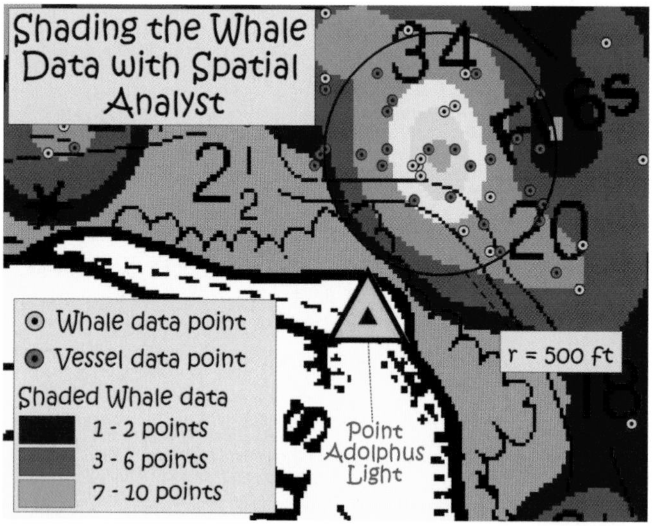

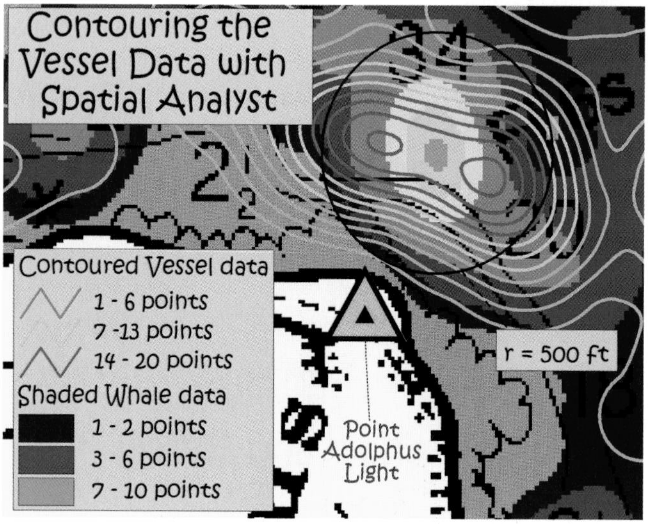

Figure 5: Use levels shown with contours.

To clearly compare two sets of point data, ArcView Spatial Analyst provides another conversion option. Note the vessel points in figure 4. Although most readers can get a sense from the point data regarding areas of high and low use, to clarify the overall pattern and make an immediate comparison with the whale data, we can apply another iteration. Much like we did with the whale point data, we can ask ArcView Spatial Analyst to calculate point density for vessel encounters. However, this time, when we select our output format, we choose to represent the spectrum of use level with contours instead of shading *(figure 5)*. Contours are derived directly from the same analytical parameters as the shading, but are effective in comparison because they can overlay shaded data without obscuring the point. If readers can understand the concepts of colored shading for one data set and contours for another, they are able to realize the comparison of the patterns of use *(figure 6)*.

The next steps in cooperative stewardship

SEAWEAD has been able to raise an awareness of the need for cooperative stewardship in local communities, conservation groups, management agencies, and businesses. Many important questions have yet to be answered. Some of these we can confidently investigate with our current data set, while for others we still have very little information. For example, we can confidently investigate use level and its relationship to bathymetry; however, we have minimal information about the acoustical effect of vessels on humpback whale behavior.

At this point, our success lies in creating a forum and opening the doors for cooperative stewardship of marine habitat at Point Adolphus. If we are going to seriously engage in long-term planning we must enter this door with wide eyes and mutual respect. More information is available on the Web site at *www.seawead.org.*

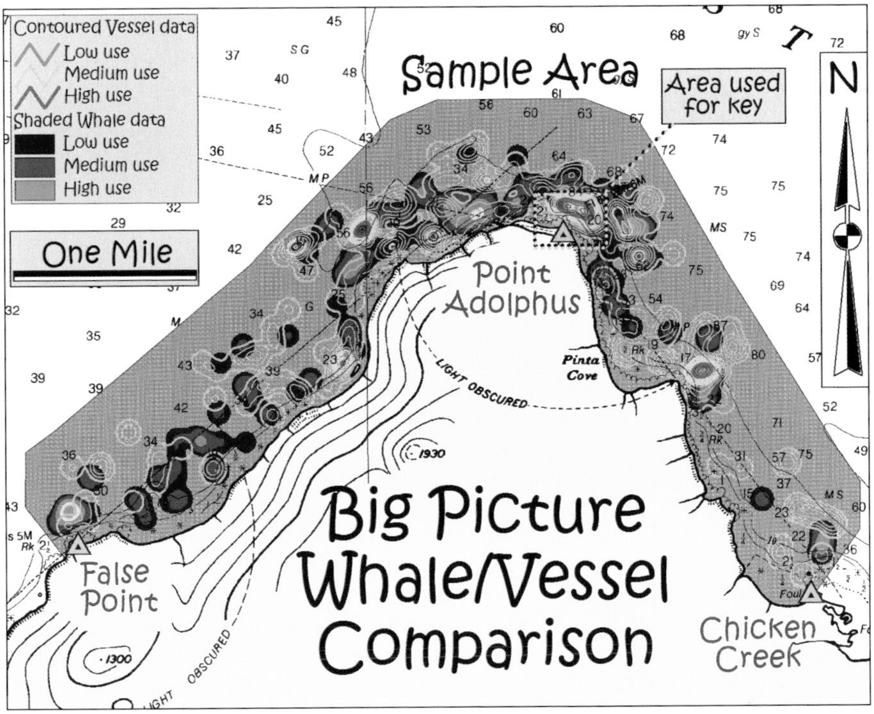

Figure 6: Overview of whale and vessel use offshore at Point Adolphus.

The "gear train" at Point Adolphus, a good example of primal cooperation.

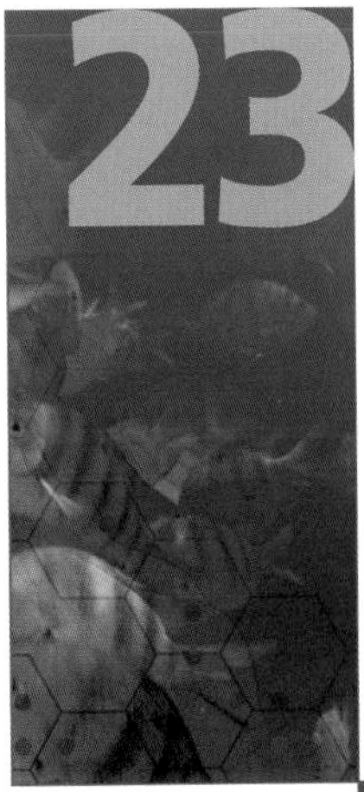

A GIS Recipe for Determining Benthic Complexity

An Indicator of Species Richness

Jeff Ardron
Living Oceans Society
Sointula, British Columbia, Canada

While conservationists agree there is a pressing need to set aside wild places, there is little agreement on how exactly to choose such places, and less still with regard to the marine environment. Of the theories and practices that do exist, most rely heavily on good biological inventories. But what if those data sets do not exist?

In this article, I present the Living Oceans Society's development of a measure of physical (benthic) complexity that is useful as an indicator of distinctive heterogeneous habitats often associated with species richness. In the absence of good comprehensive biological data, physical complexity can be used as a surrogate to direct marine planning and research activities. I will focus on how we calculated benthic (bottom) complexity. It is relatively straightforward and requires only bathymetric data—usually one of the best data sets available for marine analyses.

Background

In 1998, the Living Oceans Society took on the challenge of designing a science-based network of marine protected areas for the Central Coast of British Columbia, Canada. The coast extends approximately 425 kilometers (~265 miles) diagonally in a northwest to southeast orientation. It is characterized by a deeply crenulated shoreline (~13,000 kilometers or ~8,100 miles), including many fjords, rocky reefs, and archipelagos. The Central Coast is among the most remote in British Columbia, and consequently the least studied. This has led to a paucity of reliable biological data. However, its remoteness has ensured that it still somewhat bears the stamp of "wilderness," despite declining fish stocks and the recent expansion of salmon farms.

We believe fully protected marine reserves are a cornerstone to achieving lasting marine biodiversity and sustainable resource use. Presently, there are no fully protected areas in the Central Coast waters, and only a handful elsewhere in the province, accounting for a meager 300 hectares (~740 acres) or so. Unlike traditional "command-and-control" fisheries management restrictions, fully protected marine reserves offer an alternative philosophy: managing what we do not know. We simply don't know enough to manage each and every marine species in a traditional single-species fashion. Managing what we do not know still requires, however, that we do our best to manage as much as possible. With regard to marine reserves, what this effectively means is just leaving the species alone, unbothered by humans. Furthermore, this protection must come with the least possible cost. While cost can mean many things, it is universally associated with overall area. Thus, if in the design of a reserve network, more species are protected in a smaller area, the network is considered to be efficient. Efficient reserves are more attractive from a managerial viewpoint, and may also politically have a greater likelihood of success.

There are undoubtedly many variables and criteria to be examined in designing an efficient and effective reserve network; however, most agree that it is important to consider areas of high species richness. It was to this end that we developed our measure of benthic complexity.

Physical complexity

Areas of high species richness (that is, where there are a wide variety of species) are often associated with areas of varying habitat. The more kinds of niches available in which organisms can live will usually lead to a wider variety of organisms taking up residence. Furthermore, the complexity of habitat can interrupt predator–prey relationships that in a simpler habitat might lead to the clear dominance or near extirpation of certain species. Thus, in complex habitats, species may coexist in greater diversity where elsewhere they might not. Likewise, a greater variety of life stages may also be supported. Complex habitats may exhibit greater ecosystem resilience. The question then becomes, "How do we determine the complexity of a habitat?"

Biological communities in the three-dimensional ocean can be broadly delineated into two habitat categories: (1) pelagic (water column), including physical processes and parameters such as current, temperature, salinity, dissolved oxygen, and so forth; and (2) benthic (bottom), including depth, substrate, and the pelagic parameters as they occur on the seafloor. Most of the pelagic variables change over space and time, which makes them difficult to map, except in the most general of terms. Furthermore, the data sets are not always available. In our Central Coast analysis, we looked at a subset of physical complexity, benthic (sea-bottom) complexity. Specifically, we considered the morphological shape of the seafloor, the "lay of the land," so to speak. We fully acknowledge that this is a particular subset, and is not a panacea for modeling all forms of species richness. Because the benthic habitat harbors a much greater variety of species than the pelagic, it was considered to be the best place to begin such a modeling effort. Furthermore, it is our belief that areas of benthic complexity likely have trophic "spin-offs" benefiting pelagic species.

We defined benthic complexity by how often the slope of the sea bottom changed in a given area; that is, the density of the slope of the (exaggerated) depth. Note that this is not the same as relief, which looks at the maximum change in depth *(figure 1)*. With benthic complexity, we are interested in looking at how convoluted the bottom is, not how steep or how rough, though these both play a role. Complexity is similar but not the same as "rugosity" as is sometimes used in underwater transect surveys, whereby a chain is laid down over the terrain and its length is divided by the straight-line distance. Rugosity can be strongly influenced by a single large change in depth, while for complexity all changes are treated more equally.

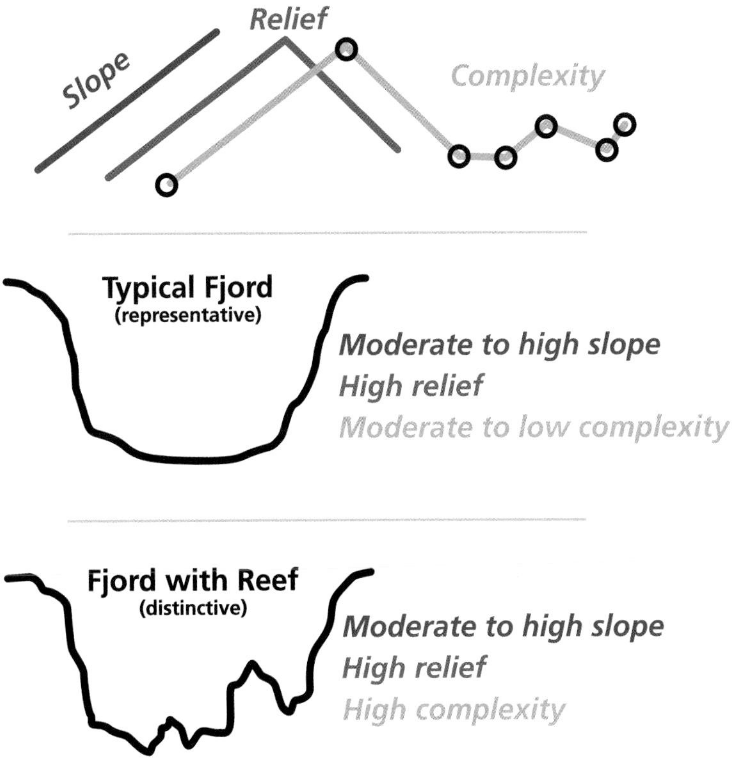

Figure 1: Slope is steepness; relief is roughness; complexity is intricacy. Complexity considers changes in slope (small circles). Complexity can distinguish typical steep-sided features such as fjords from distinctive ones, whereas measures of slope or relief generally cannot.

We created this analysis because we felt it captured biologically and physically meaningful features that the other measures missed. For example, archipelagos and rocky reefs are invariably picked out as areas of higher benthic complexity *(figure 2)*. Both are associated with several marine values. While "obvious" to the casual observer, they have no inherent quantitative definition that can be used to identify them using a GIS. Benthic complexity can identify physical features such as sills, ledges, and other distinctive habitats that are associated as biological "hot spots" providing upwellings, mixing, and refugia.

Ingredients, preparation, and tools

To perform this benthic complexity analysis, the bathymetry data must be in raster (grid) format. Such data was not available for our Central Coast analysis. We collected marine bathymetry (depth) line data from various sources and merged it into one data set, being careful to weed out redundancies, and attempting to adjust to various versions of shorelines. We then transformed the lines into evenly spaced (50-meter) points, using Bill Huber's free script, *Poly to Points* (aka *Change Lines to Points,* available on the ESRI ArcScripts page as well as Huber's own page: *www.quantdec.com/arcview1.htm*). They were then interpolated using a variety of algorithms. However, for the purposes of complexity, a triangulated irregular network (TIN)—straight linear interpolation is what we ended up using. The hard "creases" associated with TINs, which usually are troublesome, actually have certain advantages when undergoing a complexity analysis.

Figure 2: Example of benthic complexity analysis that captures a rocky reef (green oval). Bathymetric lines (gray) are in 10-meter intervals. Notice how well the convoluted lines are identified, whereas others are not, even if they are tightly packed.

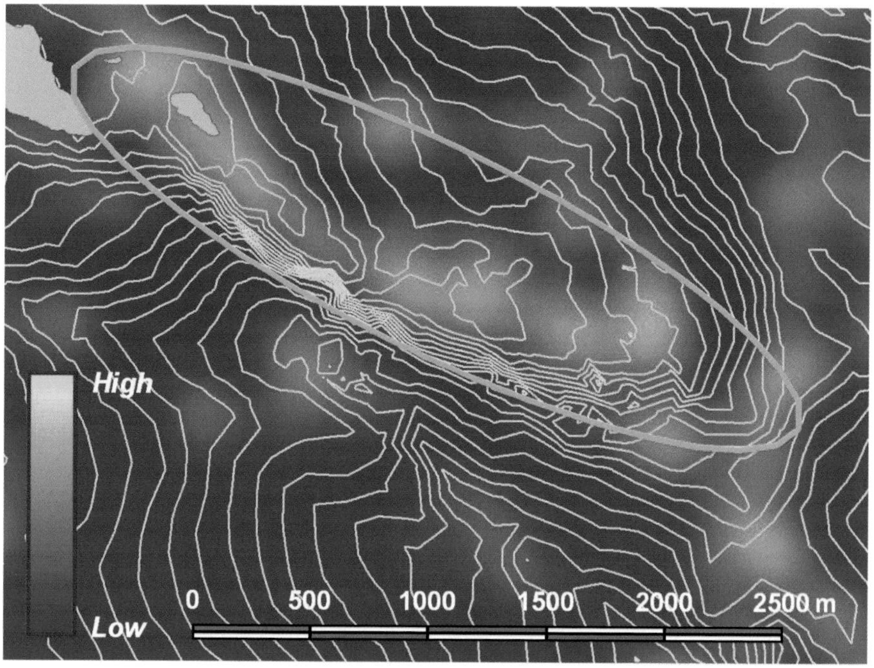

The scale and accuracy of the bathymetry data will limit the scale and accuracy of the complexity results. It is important to be aware of how evenly the depth data was sampled. Clearly, areas that are not well sampled will not show many changes in depth, and thus will not be noted as being as complex as other areas with a higher density of sampling. Often the near-shore depths are better sampled than farther offshore, which should be noted. For data sets of extremely uneven sampling, interpolations other than a TIN may be more appropriate.

We performed our analysis using ArcView 3.2 with ArcView Spatial Analyst. (ArcView 3D Analyst™ was useful but not necessary.) Other GIS with grid analysis capabilities should work too. We have performed this analysis for places other than the Central Coast, but the Central Coast study stood out because of the sheer volume of data. We used a grid size of 0.2 hectare (squares with 44.7-meter [146.7-foot] sides), which worked out to 11,369,550 cells that had a marine component. While that analysis was at a nominal scale of 1:250,000, we used the small cells to allow for certain more-detailed investigations, and to err on the side of oversampling (see step 1 in the "recipe" that follows). We performed the analysis on a 450-MHz PC with 256 MB RAM running Microsoft Windows NT®.

The recipe

The "recipe" presented below is the preferred of three different approaches we have tried. The other approaches looked at variety of roughness and variety of slope. We prefer the (density of) slope of exaggerated depth analysis due to the fewer reclassifications required, though they all produced similar results. While it might appear as if there are a number of assumptions such as exaggeration multipliers, classifications, and density radii, the analysis is quite robust. Most of the refinement of the model occurs when trying to bring out more clearly what has already shown up in a first rough trial run.

GIS recipe to model benthic complexity

1. Get the best bathymetry available and, if lines or points, convert them into a grid with the smallest cell size reasonable. Undersampling will miss important details and amplify coarse complexities beyond where they exist. This is of particular concern when studying steep-sided but narrow features such as fjords, inlets, and canyons, which can easily get cluttered. Recommendation: Cell size in meters = scale (denominator)/4000.
2. Exaggerate the depth. This will bring out any smaller changes in depth that might get overlooked. Unlike measures of relief, we are looking at all changes in depth, not just the biggest ones. This multiplier also pushes the very steep features (such as the sides of fjords) to a maximum slope approaching 90 degrees and clumps them together; otherwise, they tend to dominate the results. Recommendation: Start with 20 as a multiplier.
3. Derive the slope of the exaggerated depth.
4. Exaggerate the slope, maybe. This might require some experimentation. For small areas of detailed reliable bathymetry this is probably not necessary. For large grids, however, this exaggeration can result in whole-number output values for slope. This allows for the use of integer grids, easing memory and computational requirements. Recommendation: Do not use a multiplier to begin with.
5. Derive the slope of the (possibly exaggerated) slope. Examine the results. Look for a somewhat even spread of values, with many (two thirds, say) in the lower half of the legend. If values cluster dramatically near the low end, go back to step 4 and apply a multiplier, say, 10. If they cluster near the top, then your multiplier in step 4 is too high. If you used a TIN, you will see some of the TIN "creases." This is normal, since this is where the slope changes. Recommendation: When displayed using a graduated legend, there should be an even (log-normal) distribution of colors.

6 Reclass the results of step 5 into groupings of equal intervals. Remove the first half and keep the last half, more or less. Because we have exaggerated the depth, the lower numbers generally reflect inconsequential changes in slope as well as data variability and inconsistencies. This can be thought of like "pushing" photographic film beyond its normal ASA. The results will pick out dim differences that would otherwise go unnoticed, but will also bring out a lot of "graininess." The exact ratio will depend on the quality of the data, the exaggerations used, and a certain degree of personal judgment. Including too much leaves various unimportant artifacts; excluding too much might miss features. Look at the display: do they reflect areas that might be complex? Recommendation: Start out using 1/2.

7 Reclass the remaining grid cells into equal intervals and apply a simple weighting. For the Central Coast, we used a two-class system, rating them 1 or 2, depending on whether they were in the highest class ("2") or the second highest ("1"). I would suggest not going beyond four or five classes. Recommendation: Start out using two classes.

8 Calculate the density of these remaining cells. Calculating density is a good way to extrapolate the results to extend beyond the TIN creases, thus smoothing out the output surface. To calculate density in ArcView 3.x, the grid cells have to be converted to points. I have written a script for this, grid2pt, which is posted on the ESRI ArcScripts Web site *(arcscripts.esri.com)*. Then, calculate the density of the points. Use the class weightings in step 7 as a "population" measure; i.e., a cell with a value of 2 is worth twice that of a cell having a value of 1. The search radius will depend on the scale of your data and the sorts of questions you are asking. Bigger search radii bring out trends more clearly, but may miss details that you value. Don't be afraid to experiment using different search radii to answer different questions. I prefer using a kernel function. Recommendation: Search radius in meters = scale (denominator)/100.

9 Congratulations! Now you should have a good index of those areas that have more changes in slope than others (i.e., benthic complexity). Look at it. Chances are you'll see areas of known species richness. You may also see areas you previously had not considered. If you want to separate High from Low, then reclassify (as in step 6), keeping about the top half and tossing the lower half. You can create a shapefile from these, using three classes. To produce striking color gradients (e.g., figure 2), keep everything, but normalize the grid into a meaningful scale, such as 1–100. This will also allow for using an integer grid that will use less memory.

10 Weed out the very smallest clusters according to scale. If your study area has regional differences, it may be appropriate to buffer the remaining clusters according to the scale of the data and processes being examined.

Further reading

For more information on our MPA analysis:

Ardron, J. A., J. Lash, and D. Haggarty. 2002. Modelling a network of marine protected areas for the central coast of British Columbia. Version 3.1. Living Oceans Society, Sointula, B.C., Canada. Retrieved from *www.livingoceans.org/library.htm.*

Species richness is associated with ecosystem resilience. For a good overview, consider:

Peterson, G., C. R. Allen, and C. S. Holling. 1998. Ecological resilience, biodiversity, and scale. *Ecosystems* 1:6–18.

There is a smattering of papers discussing benthic complexity. They examine effects of bottom trawling, coral reefs, or structural (faunal) complexity:

Crowder, L., and W. Cooper. 1982. Habitat structural complexity and the interaction between bluegills and their prey. *Ecology* 63(6):1802–13.

For a brief description of a few other GIS techniques related to this topic:

Berry, J. 2000. Characterize microterrain features within data. Map analysis helps characterize microterrain features. *GeoWorld Magazine* (February, March). Internet (last accessed November 2001):

www.geoplace.com/gw/2000/0200/0200bmp.asp

www.geoplace.com/gw/2000/0300/0300bm.asp

Alaska's Marine Wildlife and Fisheries Decline

Karen Dearlove
Alaska Oceans Network

David C. Pray
Conservation GIS Center
Anchorage, Alaska

Conservation GIS Support Center
750 West Second Ave., Suite 109
Anchorage, Alaska 99501
(907) 258-6171
www.conservationgiscenter.org
A joint project of Alaska Conservation Alliance and Ecotrust

The majority of Americans, and even many Alaskans, do not know that Alaska's rich marine ecosystems are under stress and, in many cases, in decline. Moreover, they are unaware that the federal government's ability to manage these unique marine resources is compromised by conflicts of interest and an illogical division of jurisdictions. Healthy oceans are vitally important to biological diversity, species survival, and economic prosperity for rural native villages and coastal communities in Alaska. A key role of the Alaska Oceans Network is to educate the public and government officials about marine conservation issues in Alaska, and to give local and indigenous communities a voice in advocating for conservation.

The Alaska Oceans Network, a voluntary association of conservation groups, environmental groups, fishing associations, and Alaska Native organizations, seeks to restore and maintain healthy marine ecosystems in Alaska, particularly within the North Pacific. The mission of the Alaska Oceans Network is to advocate for and protect Alaska's marine ecosystems. The association unites groups with widely diverse interests for this common cause.

The Network works closely with the Conservation GIS Center located in Anchorage, Alaska. The Conservation GIS Center is a joint project of Ecotrust and Alaska Conservation Alliance (ACA). Ecotrust is an organization dedicated to supporting the emergence of a conservation economy along North America's Rainforest Coast, the region from San Francisco to Anchorage. ACA is an umbrella organization representing 45 conservation groups with more than thirty-thousand members throughout Alaska.

The Center provides GIS services, technical services, and technical equipment to both ACA member groups and other conservation-related organizations. With concentrated mapping efforts, the Center has been able to assist numerous conservation groups through education and advocacy.

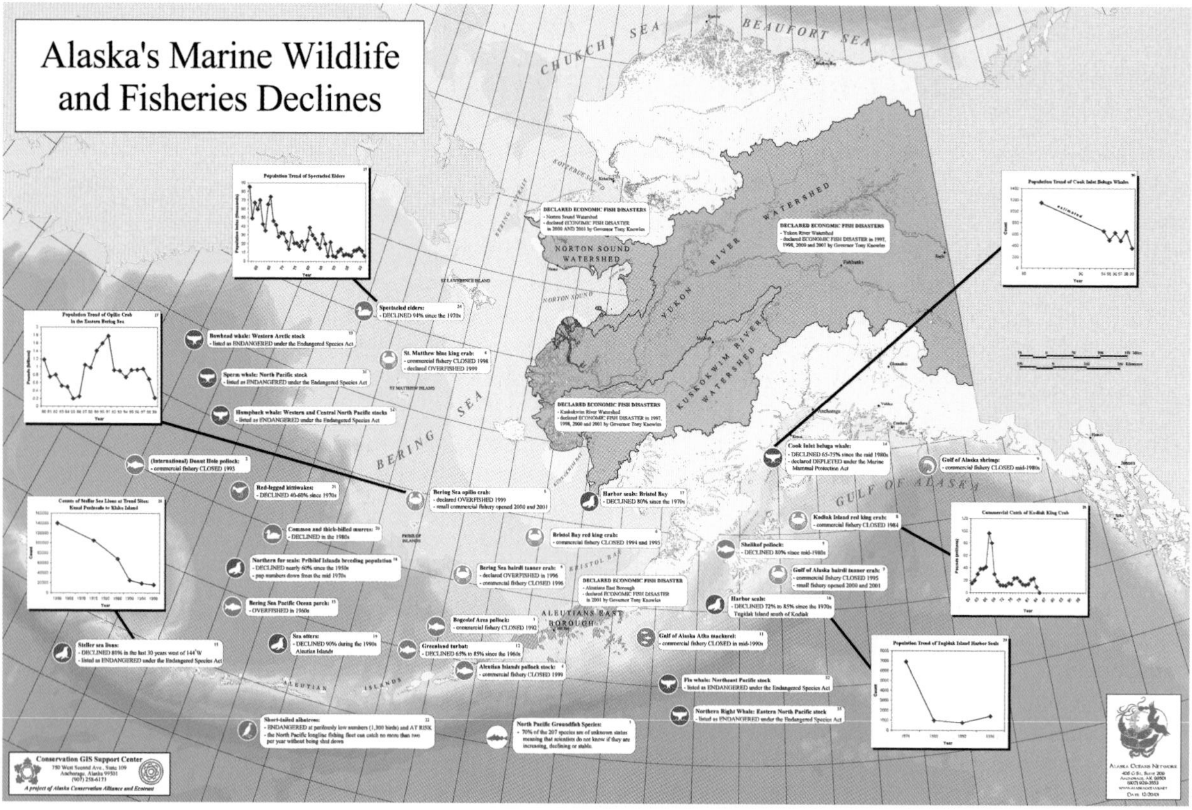

Though Alaska has one of the most ecologically diverse and pristine marine environments in America, there are several marine species in the North Pacific and the Bering Sea that are endangered, threatened, or depleted. The scientific information regarding these species and the complex marine ecosystems in which they live is either severely limited or incomplete. The map above is on the Web at *www.alaskaoceans.net/background/declinesmap.jpg.*

The Network employs a number of strategies to achieve critical reforms in fishery management policies in the North Pacific and Bering Sea, including grassroots and media outreach, scientific research, advocacy, and legal action. These activities are intended to:

- Reduce by-catch: the accidental taking of nontargeted species
- Address overfishing and its impacts on complex marine food web systems
- Protect marine habitat for both plant and animal species
- Promote sustainable, ecosystem-based fishery management practices

Now entering its second year, the Network has focused its efforts on commercial fisheries reform in the North Pacific and Bering Sea. These reform efforts have helped to develop working relationships among diverse groups. These relationships represent a solid foundation for future cooperative work on these and other complex environmental threats to Alaska's marine environment.

As part of its education effort, the Network gathers complex research information from a variety of sources and presents it in a straightforward, spatial-graphic manner. The map, *Alaska's Marine Wildlife and Fisheries Declines* (opposite page), is one example of the Network's efforts in this area. Its GIS is displaying scientific information that has never before been geographically presented. The map details several states of decline of various species in Alaska's marine environment. It also depicts a clear sense of the ecological conditions in the North Pacific and Bering Sea.

Despite these commercial and noncommercial species declines, Alaska continues to have the most productive marine ecosystems in the United States, and arguably, the world. The state's marine waters include the northern region of the Pacific Ocean, the Bering Sea, the Chukchi Sea, and the Arctic Ocean. Alaska's marine environment includes a coastline longer than that of the rest of the United States; 34,000 miles of coastline stretch from the temperate rain forests of Southeast Alaska to the arctic tundra. Alaska's outer continental shelf makes up 74 percent of the United States' total shelf area, approximately 1.5 million square miles—nine times the shelf area of California. Alaska's marine waters harbor the largest populations of marine mammals and seabirds in the Northern Hemisphere, the greatest number of wild salmon stocks in North America, and the most prolific stocks of pelagic and benthic (or bottom-feeding) fish in the world.

The phytoplankton that bloom along the edge of the melting polar ice is the source of much of the fisheries' productivity, which, in turn, feeds the seabirds and marine mammals. Alaska's marine mammal menagerie includes spotted, ribbon, bearded, ringed, northern fur, and harbor seals; Pacific walrus; Steller sea lions; sea otters; harbor and Dahl porpoise; and bowhead, fin, humpback, right, blue, sei, killer, sperm, beluga, and minke whales. The coastal and nearshore areas provide important nesting grounds for waterfowl, including 18 species of diving ducks and 10 species of geese, with more than 50 percent of the North American population of tundra swans, five species of geese, 13 species of ducks, as well as loons and sandhill cranes nesting in the region of the Yukon–Kuskokwim Delta. The Bering Sea is a particularly important region for seabirds. More than 90 percent of the seabirds found in the continental United States are also found in this region during the breeding season.

Today, these nationally and internationally critical marine ecosystems are showing significant signs of stress. Higher trophic species, such as Steller sea lions and certain types of seabirds, are listed as threatened or endangered under the Endangered Species Act (ESA). Data from the National Marine Fisheries Service (NMFS), the Alaska Department of Fish and Game (ADFG), and Governor Tony Knowles' office underscores some of these concerns:

- Steller sea lion populations have declined by 80 percent in the western Gulf of Alaska and Bering Sea, and are now listed as endangered under the ESA.
- Red-legged Kittiwakes, a member of the gull family, have declined by 77 percent in the Pribilof Islands.
- Three species of seabirds are listed as threatened or endangered: the spectacled eider, Steller's eider, and the short-tailed albatross.
- North Pacific fisheries (those including pollock, crab, and shrimp) are in decline, with several fisheries now closed in certain areas.
- In 2000 and 2001, Governor Knowles declared economic disasters for three salmon fisheries due to a lack of spawning king and chum salmon in the estuary systems of the Yukon and Kuskokwim Rivers and Norton Sound.

Each year, fishing vessels in the North Pacific haul in various amounts of an estimated 1,000 species of fish and other marine life, only to throw it back dead or dying as by-catch. Data collected by onboard observers indicates that over 300 million pounds of groundfish were discarded in 1999. Annual by-catch estimates for 1999 included 14 million pounds of halibut, 2 million pounds of herring, 97,000 individual salmon and nearly 3 million crabs. One of the worst examples is the Yellowfin Sole Fishery, a bottom trawl fishery, that threw away 151,800,000 pounds of by-catch, according to the National Marine Fisheries Service.

Recent court rulings underscore the need for federal fisheries management reforms to end wasteful and destructive commercial fishing practices that impact the marine food web and destroy marine habitat. A federal district court's 1999 ruling in Greenpeace, et al. v. NMFS, et al. (Civ. No. C98-0492Z) upheld the agency's determination that pollock fisheries were likely to jeopardize the continued existence of endangered Steller sea lions and to adversely modify their critical habitat. The court ruled that the NMFS failed to adequately analyze the impact of groundfish fisheries on the endangered Steller sea lion. Just as important, the court also ruled that NMFS failed to analyze the full impact of groundfish fisheries on the North Pacific marine ecosystem as a whole. In response to this ruling, NMFS undertook the first National Environmental Policy Act (NEPA) review of federal Fishery Management Plans (FMPs) in Alaska in nearly 20 years. This review of groundfish fisheries in Alaska provides a unique public opportunity to assess alternatives to North Pacific FMPs that ensure that an ecosystem-based framework is applied. This public process is also a vehicle for both identifying sustainable management alternatives and for holding NMFS accountable for the result.

During this NEPA review process that culminated in a Draft Programmatic Supplemental Environmental Impact Statement (SEIS), Network staff coordinated and assisted with member groups' SEIS comments; partnered with the Alaska Center for the Environment to send a letter to its 10,000 members; and garnered the support of the Partnership Project that connected with more than 200,000 environmental activists to comment on the SEIS nationwide. More than 21,361 public comments representing all 50 states and 28 foreign countries were delivered to NMFS regarding the SEIS on groundfish fisheries. Of these comments, more than 17,000 reflected the scientific concerns and positions of the Network. This response, coupled with a review by the Environmental Protection Agency, has forced NMFS to redraft the SEIS to comply with NEPA standards and to address critical scientific and public concerns over fisheries' management that includes the needs of the entire ecosystem. The SEIS will be redrafted over the next two years.

The Network's public outreach and media efforts during the SEIS process have resulted in broad public education on the issues. They have also built partnerships and expanded the marine conservation network, both within Alaska and nationally, to address federal fisheries management reform efforts. As part of its outreach, the Network produced tools such as the map *Alaska's Marine Wildlife and Fisheries Declines,* which it disseminates to members, partners, and officials at the state and federal levels. These tools illustrate the magnitude of stresses and complexities of issues that impact Alaska's marine systems. This initial effort has generated public interest, expanded public understanding, and prompted public response. Already, our effort has led to reforms in federal fisheries management. Both the Alaska Oceans Network and Conservation GIS Center hope that continued efforts will foster the restoration of endangered marine species and the protection of an extraordinary national treasure: Alaska's marine environment.

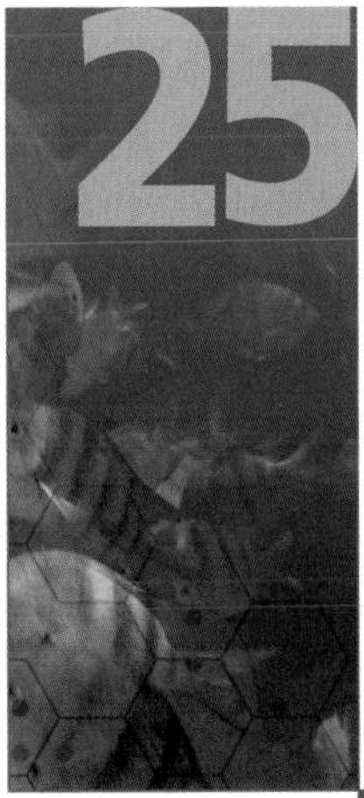

USGS Glacier Bay Field Station Develops GIS Tools

Philip N. Hooge
U.S. Geological Survey Glacier Bay Field Station
Glacier Bay National Park and Preserve, Alaska

The USGS Glacier Bay Field Station takes GIS seriously when developing tools for its own marine geographic research. The project's scientists have come up with some worthwhile GIS marine geography solutions. Although these were originally designed for Glacier Bay research projects, the designers had the foresight to make them adaptable by others in the marine geography GIS community.

One popular extension arose in an effort to perform spatial studies of movement behavior that can be integrated into the GIS. The GIS team developed software that integrates ArcView with a large collection of animal movement analysis tools. This application, Animal Movement, can be loaded as an extension under multiple operating system platforms. The extension contains more than 40 functions specifically designed to aid in the analysis of animal movement; these include parametric and nonparametric home-range analyses, random walk models, habitat analyses, point and circular statistics, tests of complete spatial randomness, tests for autocorrelation and sample size, point and line manipulation tools, and animation tools. This data could be collected from radio tags, sonic tags, ARGOS satellite tags, or observational data. The program is designed to implement a wide variety of animal movement functions in an integrated GIS environment. The program also has significant utility for analyzing other point phenomena.

Landsat false color composite of Glacier Bay.

Underwater rebreather setup for benthic mapping.

PHOTO COURTESY OF JEFF MONDRAGON

PHOTO COURTESY OF JOHN BROOKS

PHOTO COURTESY OF JOHN BROOKS

The bay from Marble Mountain (above); mountain and driftwood (below).

Tidewater glacier.

PHOTO COURTESY OF JOHN BROOKS

Oceanographic Analyst is an ArcView extension that allows three- and four-dimensional analysis and display of volumetric and time-series oceanographic data. It is the only tool available at the current time that permits integration of full-cast oceanographic data into a GIS. Its functions can also be used for calculating photic depth, integrating chlorophyll-a, processing weather data, exporting and importing, aggregation, and summarizing. Oceanographic Analyst will soon have the ability to analyze Doppler current data and visualize them as 3-D vectors animated over time.

Two other ArcView extensions on this site worth noting are Spatial Tools and the Population Viability Analyst (PVA). Spatial Tools extends the capability of ESRI's ArcView Spatial Analyst extension by allowing additional menu-based tools for raster data manipulation including creating mosaics, merging, clipping, contour profiling, cleaning, and warping, among others. PVA is a powerful method for evaluating the probability of a population's extinction in the face of variation. The PVA ArcView extension offers an alternative to traditional population models that are strictly deterministic. The PVA extension is designed to examine the influences of multiple types of variation, including differences in the values of population parameters, such as initial population sizes and birth or death rates and differences in the way those parameters vary. This extension allows the user to create life tables based on overall population parameters or to utilize a fully populated life table to conduct multiple population simulations. The program is in its development stage, and some caution is advised in interpreting results.

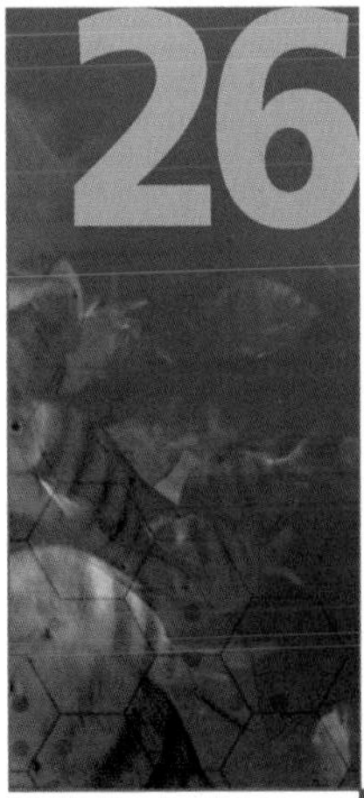

The Orca Pass International Stewardship Area Process

Where Marine Science and Policy Intersect

Philip Bloch, Mike Sato, and Jacques White
People For Puget Sound
Seattle, Washington

PEOPLE FOR PUGET SOUND
pugetsound.org

Marine fisheries have declined steeply throughout the twentieth century, and many global stocks are at overfished or depressed levels today. The worldwide magnitude of fishery losses is enormous in terms of biomass and abundances of large animals (Jackson et al. 2001). Public recognition is growing that the world's marine and aquatic ecosystems are considerably less healthy and are therefore providing fewer ecological and economic benefits (Costanza et al. 1998) than before humans began impacting these systems on a global scale. Today it is clear that contrary to views expressed by Thomas Huxley in the late 1800s, fish stocks both regionally and worldwide are not inexhaustible (Huxley 1884).

While the declines in fish stocks are notable, the distributions of marine species and fishing effort are equally important. Maps have a rich history as a navigational tool for fishermen and marine scientists. Fishing spots and reefs have long been noted on maps, and with the advent of GIS people have begun analyzing these maps for trends and to examine resource distributions. Marine managers are able to make more informed decisions today because they can understand not only how many fish are caught, but also where they are being caught. By mapping the distribution of marine species and habitats we are more able to strategically protect the marine environment.

Past technologies permitted humans to harvest marine species within a relatively narrow range of depths and distances from shore. Thus, technological limitations created and maintained areas that acted as refugia. However, with increases in human population levels, improved harvest technologies, and increased harvest efforts, many fisheries areas have begun to act as resource sinks. Such resource sinks were historically buffered by larvae and nutrients coming from nearby source populations that were naturally protected because they were too deep, too remote, too rough, or too dangerous to fish (Roberts et al. 2001). Today few natural refugia are able to avoid the impacts of resource extraction and disturbance created by humans. Without a source population providing immigrants to offset the losses resulting from harvest, many fisheries appear to be collapsing, with larger, commercially valuable species disappearing first (Jackson et al. 2001). Ultimately, overfished populations

Figure 1: The inland marine waters of Washington, including Puget Sound and the adjacent waters.

are no longer abundant enough to interact significantly with other species in the community, and species assemblages shift with new species becoming dominant (Cushing 1988).

Credible reports have documented the dramatic decline in the health and abundance of most marine species in Puget Sound *(figure 1)* and the adjacent waters (West 1997, Washington Department of Fish and Wildlife 1993). In Washington at least 13 species and species groups, including several that are not directly harvested, have undergone significant, documented declines in regional population abundance (West 1997). The causes of decline are many, including near-shore and estuarine habitat loss, water and sediment contamination, reduced abundance of prey, overharvest, and human disturbance from such things as recreational activities and marine transportation. To address these causes will require multiple strategies, including support and involvement of the public and user groups; gathering and intelligently applying scientific data; development of ecosystem-specific management actions; and monitoring, evaluation, and adjustment of the actions employed.

Significant declines are evident in harvested fishery resources, as well as quantitative catch per-unit data. This emphasizes the need for a precautionary fisheries management approach (Agardy 1994); protecting and restoring populations of marine species will require far more than simple adjustments in traditional fisheries management. Fisheries management is susceptible to social and economic pressures and complicated by periodic unanticipated natural fluctuations in the abundances of target species (Jamieson and Levings 2001). Regulating catch does little to mitigate the impacts of anthropogenic development and habitat destruction, and single-species management schemes fail to account for by-catch of nontargeted species. Traditional fisheries management should not be abandoned entirely, but it should be supplemented, particularly in those situations when it is most susceptible to failure. Unfortunately, marine conservation efforts have tended to lag behind similar terrestrial conservation efforts by roughly two decades (Agardy 1994). This gap is evidenced by the presence of conservation wilderness areas that were created to protect terrestrial parks more than a century ago, while marine protected areas (MPAs) have only become widely accepted in the last 40 years (Norse 1993).

MPAs are defined by the International Union for Conservation of Nature and Natural Resources as "any area of intertidal or subtidal terrain, together with its overlying water and associated flora, fauna, historical and cultural features, which have been reserved by law or other effective means to protect part or all of the enclosed environment" (Kelleher and Kenchington 1992). Such protected areas are growing in popularity worldwide, as it is estimated that there were 118 MPAs in 1970 (Kelleher and Kenchington 1992) and 1,306 by 1995 (Kelleher et al. 1995). These estimates do not include state, provincial, and locally established MPAs because the mechanisms for development and management are so numerous and diffuse that comprehensive identification of programs and smaller protected areas has been impossible (Murray 1998). However, in the state of Washington, 118 areas fitting a broad MPA definition have been identified (Murray 1998) and an additional 124 have been documented in British Columbia (Jamieson and Lessard 2000). Despite the apparent quantity of "MPAs" identified in Washington, conservation benefits may be grossly overestimated, as only one is an actual no-take reserve, while most (84 of 118) are primarily seaward extensions of terrestrial parks that serve primarily recreational purposes offering little, if any, resource protection.

By reserving places using MPAs, sensitive life stages may be protected, habitats are protected from incidental damage, and marine species are protected from direct and indirect harvest impacts. However, MPAs are not the only mechanism for protecting and restoring marine species and should be used as part of a portfolio of regulatory and nonregulatory tools to protect and restore fisheries, habitat, and ecosystems. MPAs are a low-risk solution and therefore an appropriate precautionary management step.

Beyond benefits for fisheries and the conservation of species and biodiversity, protected areas provide a mechanism for educating the public. Much like old-growth forests, MPAs can harbor types, sizes, and numbers of species not observed in harvested areas. These areas also have considerable value

Table 1: Sound & Straits Coalition members working toward making the Orca Pass International Stewardship Area a reality.

In British Columbia	In Washington State
Georgia Strait Alliance—*B.C. Coordinating Group*	People for Puget Sound—*Washington Coordinating Group*
Canadian Parks and Wilderness Society	Evergreen Islands
Galiano Conservancy Association	Friends of the San Juans
Islands Trust	Orca Conservancy
Living Oceans Society	Orca Network
Mayne Island Naturalists	Orca Recovery Campaign (Earth Island Institute)
Oceans Blue Foundation	San Juan County
Pender Island Conservancy	SoundWatch
Society Promoting Environmental Conservation	Surfrider Foundation (Pacific Northwest Region)
Underwater Council of BC	Washington Scuba Alliance

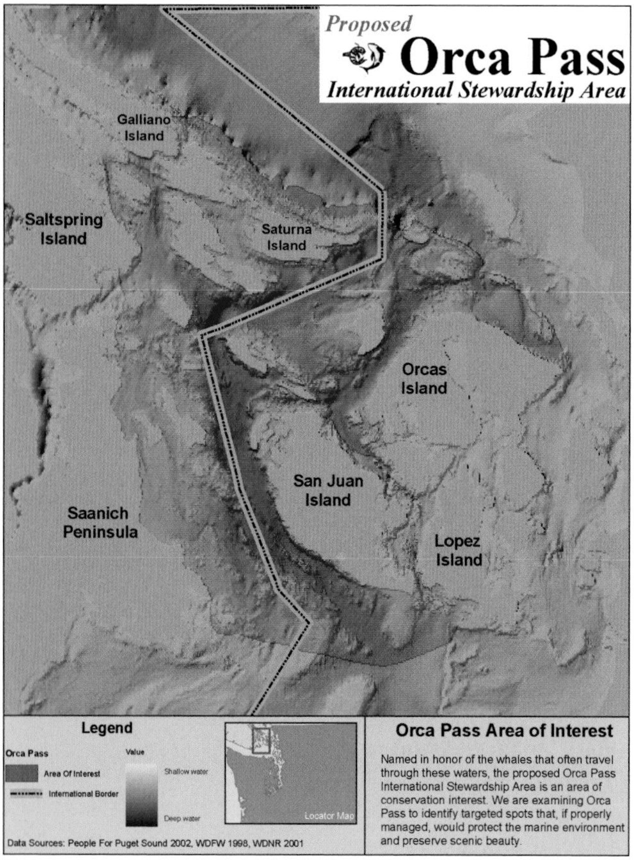

Figure 2: The Orca Pass International Stewardship Area straddles the border between the United States and Canada.

as natural laboratories. The success of MPAs should be measured using the full range of potential benefits, not just those offered to commercial fisheries.

The continued failure of traditional fishery regulations to adequately protect marine resources has prompted many organizations and individuals to call for the establishment of marine protected areas. One such effort grew out of the abortive attempts to establish a National Marine Sanctuary in the Northwest Straits. This effort focuses on what is known as the Orca Pass International Stewardship Area *(figure 2),* and is an initiative of American and Canadian conservation, commercial, and recreational groups to establish a stewardship area between Washington State's San Juan Islands and the Gulf Islands of British Columbia *(table 1).* Validating and emphasizing the need for improved resource management in this area, the effort was joined and formally endorsed in 2000 by the local governments for San Juan County and the Islands Trust.

One of the principle mechanisms proposed for protecting the transboundary region encompassed by the Orca Pass is to develop a network of MPAs. In choosing areas for protection in such a network, it is critical to identify clear goals, objectives, and expectations, such as the purpose of each reserve; the species, communities, and habitats to be protected; and the projected role and contribution of each reserve to the network (Murray et al. 1999). Through meetings with various managers, stakeholders, and interested parties, a clear purpose and role for MPAs in the region has been developed. For Orca Pass, the proposed primary goal of reserves is to protect representative samples of the local ecosystem, with an emphasis on protecting those areas of highest biodiversity and those areas used by or supporting species identified as declining or endangered. A secondary goal is that the network of protected areas buffer populations of economically valuable species from declines by protecting a portion of the known breeding grounds and protecting a portion of the adult population from harvest. Combined, these initiatives will help conserve unique and essential parts of the ecosystem by protecting those areas from most forms of anthropogenic impacts.

Having made a commitment to conserve biodiversity in the Orca Pass area, there is a need to explicitly identify which discrete locations should be protected. Due to economic and political constraints on the amount of land and water that may be set side for conservation, there is a need to identify areas that efficiently protect and ensure the continued survival of both individual species and ecosystem functions.

The first step in identifying areas to conserve is to identify what should be conserved. By involving managers and interested stakeholders in the site identification process, a consensus was developed for the intent of MPAs, which led to a logical group of target species and habitats. For Orca Pass, the conservation targets are those marine-dependent species and habitats recognized by the state and provincial governments as having a clear need for conservation through their listing on the Washington Priority Habitat and Species List (Washington Department of Fish and Wildlife 1996) and the B.C. Red and Blue lists (Conservation Data Centre 2001).

Due to the large number of agencies interested in and responsible for the management of marine species, habitats, and communities in Puget Sound and the adjacent waterways, data must be collected from a number of different sources including state, tribal, and federal resource managers, academic researchers, and nonprofit organizations. This process has been further complicated for Orca Pass because of differing policies regarding the proprietary nature of resource data. For example, critical resource data managed by the provincial government was collected by third-party contractors using funding relationships that prevent data from being released to interested parties.

While many different management agencies are ultimately responsible for parts of the local ecosystem, the primary managers responsible for fisheries and MPAs in Washington State are the state and regional tribes that co-manage fisheries. In British Columbia, a variety of provincial, federal, and tribal managers may create or enforce MPAs. Those individuals or groups ultimately responsible for establishing marine reserves are generally not scientists (Roberts and Hawkins 2000), which creates a need for transparency in any scientific approach to establishing MPAs.

Historically, marine protected areas have been designated independent of one another, or without consideration of a regional network. Most existing MPAs are small, and very few have an established management plan. Examples of designations in the region include underwater parks, special fishery management areas, preserves established by private nongovernmental organizations

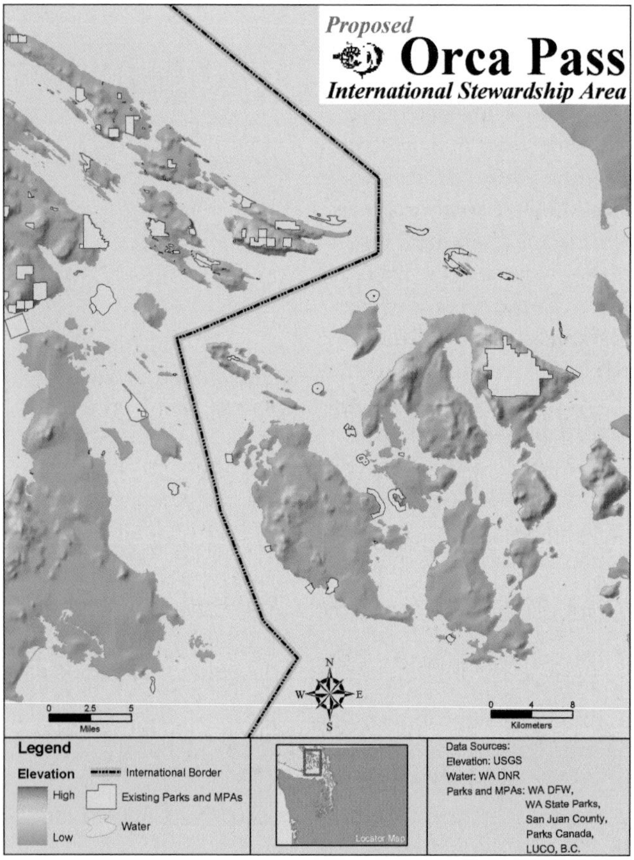

Figure 3: An overview of existing marine protected areas and parks near the Orca Pass International Stewardship Area.

(NGOs), state parks, national historic preserves and reserves, national wildlife refuges, and voluntary bottomfish recovery reserves. These existing reserves comprise a de facto MPA network *(figure 3);* therefore the Orca Pass effort is focused on identifying additional areas that contribute to protecting regional biodiversity to fill gaps in the existing MPA network.

The overriding goal for Orca Pass is to create a zoning scheme that would create not only no-take areas that might serve as the core for a network of MPAs, but also transition zones where some types of harvest might be regulated while others would not. By using a zoning framework we hope to facilitate cooperation among the Canadian and American resource managers. Sites identified for protection may become regulatory no-take areas adopted by the state department of fish and wildlife, or they may become voluntary refuges adopted by and policed by the public to restrict harvest. Ultimately we hope to meaningfully engage the public in discussions about the stewardship and protection of our marine resources. We have already begun this process by encouraging scientists and resource experts to identify sites and species of interest. Working toward making Orca Pass a reality, the coalition will continue its consultations with scientists and resource stewards, the general public and decision makers, and constituencies (kayakers, whale watchers and whale-watch tour operators, and scuba divers) who might be encouraged to support protection of marine resources.

Before we can understand how to effectively protect species and their resources, it is important to understand their spatial distribution and abundance. Today, resource managers in Washington and British Columbia use geographic information systems (GIS) to map most of their biological data collected for Puget Sound and the adjacent waters. People For Puget Sound has worked to augment this information by bringing resource experts together in workshops to supplement and critique existing data sets. These critiques have identified several overriding limitations to working with a diversity of species and data sets. First, species and habitat distribution data has been collected using a variety of survey methodologies, and often spatially incomplete surveys represent the best available information. Additionally, data is often collected from surveys designed for a specific purpose other than assessing the distribution and abundance of a given species or habitat. For example, data that is compiled by fishery management units that may or may not correspond to biologically meaningful boundaries may be less valuable than complete surveys. To accommodate the variety of data types uncovered for Orca Pass, spatial species distribution data is reclassified into one of three classes: locations where species were observed (confirmed present); locations that have never been surveyed or are of unknown survey status (unknown status); and locations that have been surveyed and species were not observed (presumed absent).

Once species and habitat distribution information is in a common currency we can begin doing multispecies comparisons. The simplest of these is to calculate the species richness for a given site. GIS facilitates the combining or layering of different data sets to integrate the amount of resources found at a given location. For Orca Pass we chose to subdivide the area of interest into 25-hectare subsections for analytical purposes. This size analytical unit facilitates analysis of distributions at approximately the same scale that many of the target species interact with their habitat. By counting the number of species whose distribution fell within each analytical unit we were able to identify several areas that appear to have high species richness.

While representing biological "hot spots" for species of interest is valuable from a planning perspective *(figure 4),* it falls far short of our goal of identifying discrete locations that might serve as core MPAs. Two additional methods are being employed to help analyze sites for conservation. First, we are using a heuristic algorithm that was originally developed by Hugh Possingham and Ian Ball to identify areas that maximize species representation while minimizing the area included in reserves. This algorithm selects the set of locations that achieves minimum goals of species or habitat representation while trying to minimize the amount of habitat being selected. These methods identify a network of areas that meet species representation goals with the fewest possible locations.

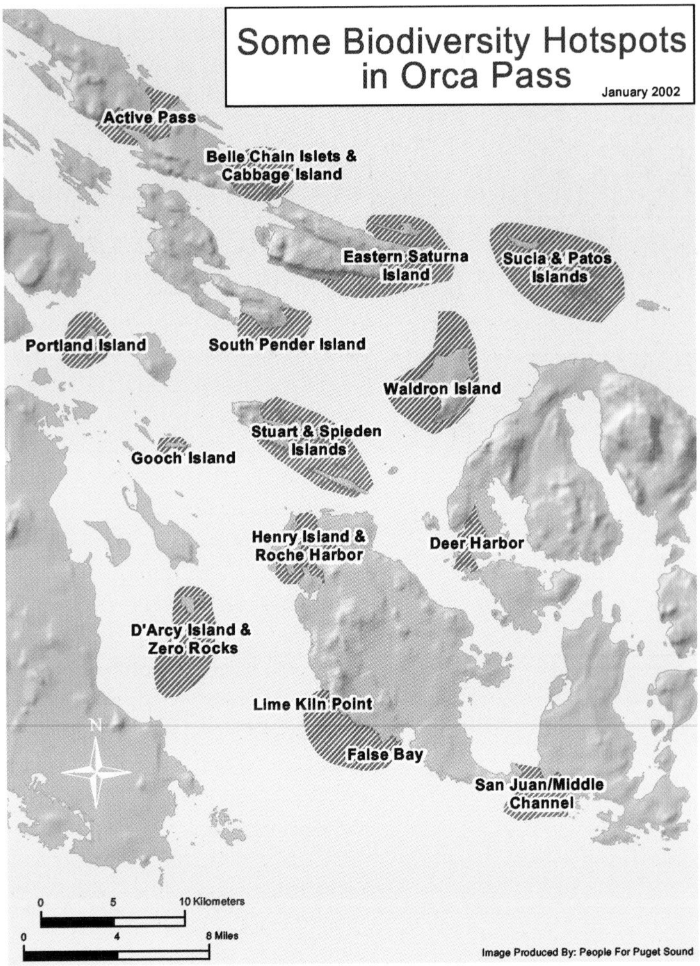

Figure 4: Some biodiversity hot spots identified by examining species distributions.

Figure 5: An example of a graph with six nodes (a, b, c, d, e, f) representing potential reserve sites, and edges representing potential linkages between sites.

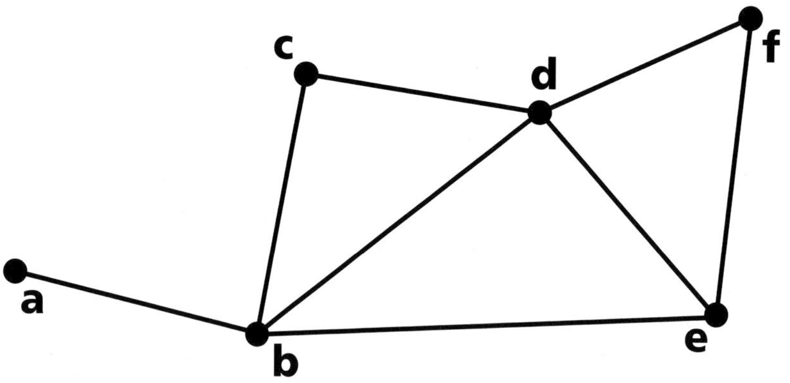

The second tool that we are applying is a graph theoretical approach (Urban and Keitt 2001) that will allow us to account for linkages between proposed areas *(figure 5)*. This will allow us to explicitly analyze one of the fundamental principles of designing marine protected areas: that effective MPAs must be self-sustaining and capable of supplying larvae to other protected and unprotected areas (e.g., Carr and Reed 1993). This approach starts with the assumption that sites selected using other methods are potential reserves and serve as nodes in our network, and examines connections between individual sites. This tool will enable us to identify those patches whose loss would most seriously affect the overall connectivity of the network (e.g., Bunn et al. 2000).

Despite the clear benefits of MPA designation, this mode of conservation is relatively new. Changing the approach to conserving marine species is a slow process and the burden of proof falls on the new approach even though the existing approach has clear imperfections. However, with growing interest in MPAs, many stakeholders have become more active in advocating and understanding mechanisms of marine protection. Free and open communication is critical to fostering any MPA effort. Sites are ultimately identified for conservation because of the stated conservation and protection goals. To accommodate economic and social interests, there is a need to display alternative protection schemes that clearly delineate which goals are fully reached and which are not being met.

Raising the awareness of resource users, managers, and the general public of the need to conserve, restore, and protect marine resources is only the beginning. The Orca Pass effort is unique in that it is attempting to engage a wide range of managers and stakeholders in identifying places that should be set aside as MPAs. MPAs represent a rare opportunity for a regulatory tool to function as an educational mechanism by protecting areas for future generations to observe, resolving user conflicts, and creating natural laboratories for research and exploration.

References

Agardy, T. 1994. Advances in marine conservation: The role of marine protected areas. *Trends in Ecology and Evolution* 7:267–70.

Bunn, A. G., D. L. Urban, and T. H. Keitt. 2000. Landscape connectivity: A conservation application of graph theory. *Journal of Environmental Management* 59:265–78.

Carr, M. H., and D. C. Reed. 1993. Conceptual issues relevant to marine harvest refuges: Examples from temperate reef fishes. *Canadian Journal of Aquatic Science* 50:2019–28.

Conservation Data Centre (CDC). 2001. *Provincial red and blue list.* British Columbia Conservation Data Centre.

Costanza, R., F. Andrade, P. Antunes, M. van den Belt, D. Boersma, D. F. Boesch, F. Catarino, S. Hanna, K. Limburg, B. Low, M. Molitor, J. G. Pereira, S. Rayner, and R. Santos. 1998. Principles for sustainable governance of the oceans. *Science* 281:198–99.

Cushing, D. H. 1988. *The provident sea.* Cambridge: Cambridge University Press.

Huxley, T. H. 1884. Fisheries exhibition literature 4, inaugural address.

Jackson, J. B. C., M. X. Kirby, W. H. Berger, K. A. Bjorndal, L. W. Botsford, B. J. Bourque, R. H. Bradbury, R. Cooke, J. Erlandson, J. A. Estes, T. P. Hughes, S. Kidwell, C. B. Lange, H. S. Lenihan, J. M. Pandolfi, C. H. Peterson, R. S. Steneck, M. J. Tegner, and R. R. Warner. 2001. Historical overfishing and the recent collapse of coastal ecosystems. *Science* 293:629–7.

Jamieson, G. S., and J. Lessard. 2000. *Marine protected areas and fishery closures in British Columbia.* Canadian Special Publication on Fisheries and Aquatic Science 131.

Jamieson, G. S., and C. O. Levings. 2001. Marine protected areas in Canada: Implications for both conservation and fisheries management. *Canadian Journal of Fisheries and Aquatic Science* 58:138–56.

Kelleher, G., and R. Kenchington. 1992. Guidelines for establishing marine protected areas. *A marine conservation and development report.* Gland, Switzerland: IUCN.

Kelleher, G., G. Bleakley, and S. Wells., eds. 1995. *A global representative system of marine protected areas. Vol. I: Antarctic, arctic, Mediterranean, northwest Atlantic, northeast Atlantic and Baltic.* The Great Barrier Reef Marine Park Authority, The World Bank and The World Conservation Union (IUCN).

Murray, M. R. 1998. The status of marine protected areas in Puget Sound. Vol. 1. Prepared for the Puget Sound/Georgia Basin International Task Force Work Group on Marine Protected Areas. Olympia, Wash.

Murray, S. N., R. F. Ambrose, J. A. Bohnsack, L. W. Botsford, M. H. Carr, G. E. Davis, P. K. Dayton, D. Gotshall, D. R. Gunderson, M. A. Hixon, J. Lubchenco, M. Mangel, A. MacCall, D. A. McArdle, J. C. Ogden, J. Roughgarden, R. M. Starr, M. J. Tegner, and M. M. Yoklavich. 1999. No-take reserve networks: Sustaining fishery populations and marine ecosystems. *Fisheries* 24(11):11–25.

Norse, E. A., ed. 1993. *Global marine biological diversity: A strategy for building conservation into decision making.* Washington, D.C.: Island Press.

Possinghan, H., I. Ball, S. Andelman. 2000. Mathematical methods for identifying representative reserve networks. In *Quantitative methods for conservation biology,* edited by S. Ferson and M. Burgman. New York: Springer-Verlag.

Roberts, C. M., B. Halpern, S. R. Palumbi, and R. R. Warner. 2001. Designing marine reserve networks: Why small isolated protected areas are not enough. *Conservation Biology In Practice* 2(3)11–17.

Roberts, C. M., and J. P. Hawkins. 2000. *Fully-protected marine reserves: A guide.* WWF Endangered Species Campaign, Washington, D.C., and University of York, U.K.

Urban, D. L., and T. H. Keitt. 2001. Landscape connectivity: A graph-theoretic perspective. *Ecology* 82(5):1205–18.

Washington Department of Fish and Wildlife (WDFW). 1996. *Priority habitats and species List.* WDFW Habitat Program.

Washington Department of Fisheries, Washington Department of Wildlife, and Western Washington Treaty Tribes. 1993. 1992 Washington State Salmon and Steelhead Stock Inventory. Olympia, Wash.

West, J. E. 1997. *Protection and restoration of marine life in the inland marine waters of Washington State.* Report prepared for the Puget Sound/Georgia Basin International Task Force, Washington Work Group on Protecting Marine Life.

Contact information

Mark Abramson
Malibu Creek Watershed Stream Team Manager
Heal the Bay
3220 Nebraska Avenue
Santa Monica, CA 90404
Phone: (310) 453-0395, ext. 146
Fax: (310) 453-7927
E-mail: streamteam@healthebay.org
Web: www.healthebay.org/streamteamhome.asp

December sunset from Point Lobos, along the California coast

PHOTO COURTESY OF MIKE OWENS

Jeff Ardron
GIS Manager
Living Oceans Society
P. O. Box 755
Salt Spring Island, BC
V8K 2W3, Canada
Phone: (250) 653-9219
E-mail: jardron@livingoceans.org
Web: www.livingoceans.org

Jerald S. Ault
Rosenstiel School of Marine and Atmospheric Science, University of Miami
Division of Marine Biology and Fisheries
4600 Rickenbacker Causeway
Miami, FL 33149
Phone: (305) 361-4881
Fax: (305) 361-4791
E-mail: jault@rsmas.miami.edu

Richard Bates
School of Geography and Geosciences
University of St. Andrews
St. Andrews
Fife, Scotland
Phone: +44 1334 463997
Fax: +44 1334 4633949
E-mail: crb@st-and.ac.uk

Tracey Baxter
Systems and Environmental Coordinator
Australian Maritime Safety Authority
GPO Box 2181
Canberra, ACT 2601, Australia
Phone: +61 2 6279 5952
E-mail: Tracey.Baxter@amsa.gov.au
Web: www.amsa.gov.au

Philip Bloch
Landscape Ecologist and GIS Manager
People For Puget Sound
911 Western Ave, Suite 580
Seattle, WA 98004
Phone: (206) 382-7007
Fax: (206) 382-7006
E-mail: pbloch@pugetsound.org
Web: www.pugetsound.org

Joe Breman
Marine and Coastal Community Manager
Environmental Systems Research Institute
380 New York Street
Redlands, CA 92373
Phone: (909) 793-2853
E-mail: jbreman@esri.com
Web: www.esri.com/industries/marine/index.html

Robert V. Burne
Department of Geology
The Australian National University
Canberra, ACT 0200, Australia
Phone: +61 26612059 / +61 25206
Fax: +61 26155544
E-mail: rburne@geology.anu.edu.au
Web: www.geology.anu.edu.au

Bob Christensen
Lead Naturalist
Southeast Alaska Wilderness Exploration, Analysis and Discovery
418 Harris Street Room 210
Juneau, AK 99801
Phone: (907) 586-3393
E-mail: bob@seawead.org
Web: www.seawead.org

Mitchell Colgan
Associate Professor and Chair
Department of Geology and Environmental Geosciences
College of Charleston
66 George Street
Charleston, SC 29424-0001
Phone: (843) 953-7171
Fax: (843) 953-5446
E-mail: colganm@cofc.edu

Karen Dearlove
Program Director
Alaska Oceans Network
406 G Street
Anchorage, AK 99501
Phone: (907) 929-3553
E-mail: karen_dearlove@alaskaoceans.net
Web: www.alaskaoceans.net

Laura K. Engelby
Dolphin Ecology Project
P. O. Box 1142
Key Largo, FL 33037
Phone: (843) 887-3422
E-mail: dolphineco@aol.com
Web: www.dolphinecology.org

Peter Etnoyer
Staff Scientist
Marine Conservation Biology Institute
15806 NE 47th Court
Redmond, WA 98052
Phone: (425) 883-8914
Fax: (425) 883-3017
Web: www.mcbi.org

Zach Ferdaña
GIS Analyst
The Nature Conservancy of Washington
217 Pine Street, Suite 1100
Seattle, WA 98101
Phone: (206) 343-4345
Fax: (206) 343-5608
E-mail: zferdana@tnc.org
Web: www.nature.org

Erik C. Franklin
Rosenstiel School of Marine and Atmospheric Science, University of Miami
Division of Marine Biology and Fisheries
4600 Rickenbacker Causeway
Miami, FL 33149
Phone: (305) 361-4881
Fax: (305) 361-4791
E-mail: efranklin@rsmas.miami.edu

Trevor Gilbert
Principal Adviser Scientific and Environmental
Australian Maritime Safety Authority
GPO Box 2181
Canberra, ACT 2601, Australia
Phone: +61 2 6279 5680
E-mail: Trevor.Gilbert@amsa.gov.au
Web: www.amsa.gov.au

Roger Goldsmith
Research/Information Systems Specialist
Woods Hole Oceanographic Institution
WHOI 161A Clark, MS 46
Woods Hole, MA 02543
Phone: (508) 289-2770
E-mail: rgoldsmith@whoi.edu

Patrick N. Halpin
Assistant Professor of the Practice of Landscape Ecology
Nicholas School of the Environment and Earth Sciences
Duke University, Box 90328
Durham, NC 27708-0328
Phone: (919) 613-8062
Fax: (919) 684-8741
E-mail: phalpin@duke.edu
Web: www.env.duke.edu/faculty/bios/halpin.html

Philip N. Hooge
USGS Glacier Bay Field Station
P. O. Box 140
Gustavuse, AK 99826
Phone: (907) 697-2230
Fax: (907) 697-2654
E-mail: Philip_hooge@nps.gov
Web: www.absc.usgs.gov/glba

Ben James
Scottish Natural Heritage
2-5 Anderson Place
Edinburgh, Scotland
Phone: +44 131 4474784
Fax: +44 131 4462405
E-mail: Ben.james@snh.gov.uk

PHOTO COURTESY OF KIM AVERY

A view from Cypress Grove Trail at Point Lobos

Martin Kaye
Bay of Fundy Marine Resource Centre
P. O. Box 273
Cornwallis Park
Nova Scotia BOS 1HO
Phone: (902) 638-3044
Fax: (902) 638-3284
E-mail: Martink@bfmrc.ns.ca
Web: www.bfmrc.ns.ca

Stuart Kininmonth
GIS Specialist
Australian Institute of Marine Science
PMB #3, Townsville Mail Centre
Townsville, Queensland, Australia 4810
Phone: + 61 7 47534334
E-mail: s.kininmonth@aims.gov.au
Web: www.aims.gov.au

Michelle Rene Kinzel
Researcher, Oceanic Resource Foundation
1095 Calle Mesita
Bonita, CA 91902
Phone: (619) 251-5484
E-mail: gypsea33@msn.com
Web: www.orf.org

James Alan Reade McClean
Research Assistant
Florida Geological Survey
Gunter Building MS# 720
903 W. Tennessee Street
Tallahassee, FL 32304-7700
Phone: (850) 488-9380
Fax: (850) 488-8086
Email: James.McClean@dep.state.fl.us

Bernd Meissner
Professor of Cartography and Remote Sensing
University of Applied Science Berlin
Luxemburger Strasse 10
Berlin, Germany, D-13353
Phone: +49 (0)30-4504-2606
Fax: +49 (0)30-4504-2632
E-mail: meissner@tfh-berlin.de
Web: www.tfh-berlin.de

Lance Morgan
Chief Scientist
Marine Conservation Biology Institute
15806 NE 47th Court
Redmond, WA 98052
Phone: (425) 883-8914
Fax: (425) 883-3017
Web: www.mcbi.org

Matthias Mueller
Graduate engineer in Cartography
Federal Maritime and Hydrographic Agency of Germany
Dierkower Damm 45
Rostock, Germany, D-18146
Phone: +49 (0)381-4563-988
E-mail: Matthias.mueller@bsh.de
Web: www.bsh.de

Chad Nelsen
Environmental Director
Surfrider Foundation
P. O. Box 6010
San Clemente, CA 92674-6010
Phone: (949) 492-8170
E-mail: cnelsen@surfrider.org
Web: www.surfrider.org

Henry Norris
Florida Marine Research Institute
Florida Fish and Wildlife Conservation Commission
100 8th Ave. SE
St. Petersburg, FL 33701
Phone: (727) 896-8626
E-mail: Henry.Norris@fwc.state.fl.us
Web: www.floridamarine.org

Joyce Palacol
International Marinelife Alliance
83 West Capitol Drive
Bo. Kapitolyo, Pasig City
1601 Philippines
Phone: (632) 638-7118
E-mail: j_palacol@hotmail.com

Christian A. Parvey
Department of Geology
The Australian National University
Canberra, ACT 0200, Australia
Phone: +61 26612059 / +61 25206
Fax: +61 26155544
E-mail: chris.parvey@anu.edu.au
Web: www.geology.anu.edu.au

Albert J. Plueddemann
Associate Scientist
Woods Hole Oceanographic Institution
WHOI, 202A Clark, MS29
Woods Hole, MA 02543
Phone: (508) 289-2789
E-mail: aplueddemann@whoi.edu

David Pray
GIS Analyst
Conservation GIS Center
750 West 2nd Ave., Suite 109
Anchorage, AK 99501
Phone: (907) 258-6173
E-mail: david@akvoice.org
Web: www.conservationgiscenter.org

Looking south down the California coast

PHOTO COURTESY OF KIM AVERY

Mark Rauscher
Environmental Programs Manager
Surfrider Foundation
P. O. Box 6010
San Clemente, CA 92674-6010
Phone: (949) 492-8170
E-mail: mrauscher@surfrider.org
Web: www.surfrider.org

Peter Rubec
Senior Research Scientist
International Marinelife Alliance
2800 4th Street North, Suite 123
St. Petersburg, FL 33704
Phone: (727) 896-8626
E-mail: peterrubec@cs.com
Web: www.marine.org

Mike Sato
North Sound Director
People For Puget Sound
911 Western Ave, Suite 580
Seattle, WA 98004
Phone: (206) 382-7007
Fax: (206) 382-7006
E-mail: msato@pugetsound.org
Web: www.pugetsound.org

Robert S. Schick
Research Associate
NOAA Fisheries, Santa Cruz Lab
110 Shaffer Road
Santa Cruz, CA 95060
E-mail: robert.schick@noaa.gov
Phone: (831) 420-3960

Steven G. Smith
Rosenstiel School of Marine and Atmospheric Science, University of Miami
Division of Marine Biology and Fisheries
4600 Rickenbacker Causeway
Miami, FL 33149
Phone: (305) 361-4881
Fax: (305) 361-4791
E-mail: sgsmith@rsmas.miami.edu

Jack Sobel
Director, Ecosystem Programs
The Ocean Conservancy
1725 DeSales St., N.W., Suite 600
Washington, D.C. 20036
Phone: (202) 429-5609
Fax: (202) 872-0619
E-mail: jsobel@oceanconservancy.org
Web: www.oceanconservancy.org

Eric A. Treml
Ph.D. student
Duke University Landscape Ecology Lab
Nicholas School of the Environment and Earth Sciences
Box 90328, Duke University
Durham, NC 27708-0328
Phone: (919) 613-8124
Fax: (919) 684-8741
E-mail: eat4@duke.edu
Web: www.env.duke.edu/landscape

Wilhelm Weinrebe
Head of Computer Center
GEOMAR Research Center for Marine Geoscience Kiel
Wischhofstrasse 1-3
Kiel, Germany, D-24148
Phone: +49 (0)431-600-2281
E-mail: wweinrebe@geomar.de
Web: www.geomar.de

Jacques White
Director of Science and Habitat Programs
People For Puget Sound
911 Western Ave, Suite 580
Seattle, WA 98004
Phone: (206) 382-7007
Fax: (206) 382-7006
E-mail: jwhite@pugetsound.org
Web: www.pugetsound.org

Doug Wilder
National Park Service
Inventory & Monitoring Data Manager
Central Alaska Network
201 1st Avenue
Fairbanks, AK 99701
Phone: (907) 455-0661
E-mail: Doug_Wilder@nps.gov

Damon J. Wing
Programs Director
Wishtoyo Foundation/Ventura CoastKeeper
3600 South Harbor Boulevard, Suite 218
Oxnard, CA 93035
Phone: (805) 382-4540
Fax: (805) 382-4541
E-mail: vck@wishtoyo.org

Dawn Wright
Associate Professor
Department of Geosciences
Oregon State University
Corvallis, OR 97331-5506
Phone: (541) 737-1229
E-mail: dawn@dusk.geo.orst.edu
Web: dusk.geo.orst.edu

Society for Conservation GIS Membership Application Form

Name: ______________________________

first name, middle initial, family name

Your Organization's Name: ______________________________

Organization Address: ______________________________

(street, city, state, ZIP/Postal Code, country)

Work Phone Number: ______________ *Work Fax Number:* ______________

Work E-mail Address: ______________________________

Organization Web Site: ______________________________

Home Address: ______________________________

(street, city, state, ZIP/Postal Code, country)

Home Phone Number: ______________ *Home Fax Number:* ______________

Home E-mail: ______________________________

Alternate Web Site: ______________________________

For more information, please visit the SCGIS Web site at *www.SCGIS.org*

Marine Geography: GIS for the Oceans and Seas
Edited by Joe Breman
Editorial assistance by David Boyles
Book design, production, and image editing by Jennifer Johnston
Copyediting by Michael Hyatt and Tiffany Wilkerson
Cover design by Steve Pablo
Cartographic assistance by Edith M. Punt
Printing coordination by Cliff Crabbe